Myelin Magic: The Remarkable Cells Shaping Our Senses

Elena

Table of Contents

Introduction

Chapter 1: The TSC1-mTOR-PLK Axis Regulates the Homeostatic Switch from Schwann Cell Proliferation to Myelination in a Stage-Specific Manner 15

Chapter 2: Glioblastoma Genetic Drivers Dictate the Function of Tumor-Associated Macrophages/Microglia and Responses to CSF1R Inhibition 36

Chapter 3: Proton Therapy Induces Remodeling of Tumor Cells and the Microenvironment in Glioblastoma 76

Chapter 4: Discussion 81

INTRODUCTION

Part I: The Microenvironment in Schwann Cell Development

<u>Schwann Cell Function in the Peripheral Nervous System</u>

Myelination of axons is vital for rapid saltatory conduction of action potentials and effective sensorimotor integration. In the peripheral nervous system (PNS), myelination is performed by a cell type known as a Schwann cell. For large diameter axons transmitting proprioception and touch, Schwann cells will form tightly compacted myelin wrappings around axon segments to facilitate rapid conduction of action potentials. In contrast, smaller caliber fibers transmitting pain and heat sensation are not myelinated but are instead ensheathed by a non-myelinating Schwann cell subset in a structure known as a Remak bundle. Both subsets of Schwann cells provide metabolic support to their associated axon. Following nerve injury, Schwann cells differentiate into repair Schwann cells and promote peripheral nerve regeneration.

Disturbances in Schwann cell development and function lead to peripheral nerve disease. Decreases in Schwann cell numbers and myelination lead to peripheral neuropathies such as Charcot-Marie-Tooth disease, Guillain-Barré Syndrome and congenital hypomyelinating neuropathy[1-4]. On the other hand, myelin overgrowth in the peripheral nervous system has also been linked to sensory neuropathies[5]. In the same vein, loss of non-myelinating Schwann cells leads to death of associated small-caliber pain and temperature axons, leading to deficits of sensory sensation in mice[6]. Increases in Remak bundle numbers from modulation of neurotransmitter signaling on non-myelinating Schwann cells are associated with allodynia and neuropathic pain in mouse models[7,8]. Understanding the mechanisms by which Schwann cells develop may offer insight into treatments of hypomyelination or hypermyelination of peripheral nerves.

<u>The Microenvironment in Schwann Cell Development</u>

Schwann cells development begins when trunk neural crest cells migrate out to the periphery, associate with embryonic nerves and differentiate into Schwann cell precursors around mouse embryonic day 12 and 13 (E12/E13) in the mouse (8-9 weeks post-conception in humans)[9]. Schwann cell precursors differentiate into immature Schwann cells between E13-E15. Starting around E15/E16 (12-14 weeks post-conception in humans) and continuing into the early postnatal period, immature Schwann cells destined to become myelinating Schwann cells undergo radial sorting and associate

1:1 with axons[10,11]. They begin synthesizing a basal lamina and differentiate into Oct6+ pro-myelin Schwann cells[10]. Both pro-myelin and immature Schwann cells proliferate to expand the total number of Schwann cells[10]. In response to microenvironmental and internal cues, pro-myelin Schwann cells commit to differentiation, decrease their expression Sox2 and Oct6, and become mature Krox20+ myelinating Schwann cells[12].

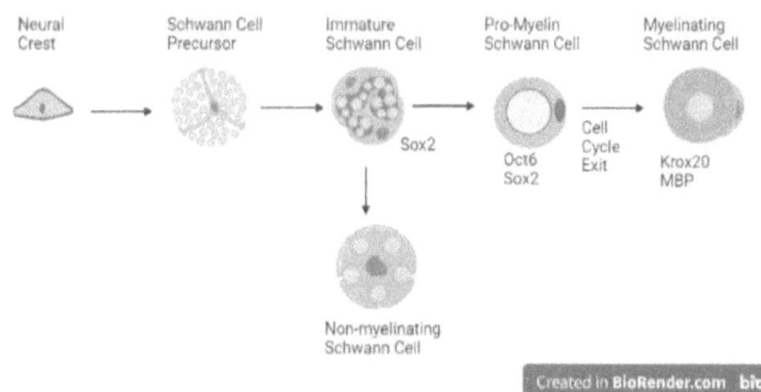

Figure 1. Schwann Cell Development. Schwann Cells are derived from the neural crest and differentiate into myelinating or non-myelinating Schwann cells.

The transitions that guide Schwann cell development are regulated by microenvironmental signals. One such source of signals are the peripheral axons with which Schwann cells are associated. Of these, type III neuregulin is the most well studied. Neuregulin binds to receptor tyrosine kinases ErbB2 and ErbB3 on the Schwann cell surface, and acts at multiple points in Schwann cell development. In early development, neuregulin signaling is required for the survival of Schwann cell precursors and stimulates the proliferation of immature Schwann cells [13,14]. During the transition from immature to myelinating Schwann cells, axonal neuregulin 1 type III is required for myelination, and the levels of neuregulin 1 correspond to the degree of myelination of the axon[15,16]. Axons also signal to Schwann cells through Jagged-Notch signaling to promote differentiation of Schwann cell precursors to immature Schwann cells, and stimulate the proliferation of immature Schwann cells[14,17]. Another axonal signal, Adam22, which binds to Schwann cell Lgi4, is required for advancement of Schwann cells beyond the pro-myelin stage[18,19].

In addition to signals from axons, developing Schwann cells also receive signals from the basement membrane and the local microenvironment. During early Schwann cell development, laminin from the basal lamina acts as a mitogenic signal by regulating GPR126-cAMP signaling[20-22]. During differentiation of immature Schwann cells to myelinating Schwann cells, basement membrane laminin is also required for appropriate differentiation of Schwann cells[23,24]. Metabolites such as glutamate and lactate also regulate Schwann cell myelination. Glutamate signaling through mGluR2 in proliferating Schwann cells also acts to promote Schwann cell proliferation and inhibit Schwann cell differentiation[25]. Transport of lactate through MCT1 into Schwann cells has been shown to be required for myelin maintenance in sensory nerves[26].

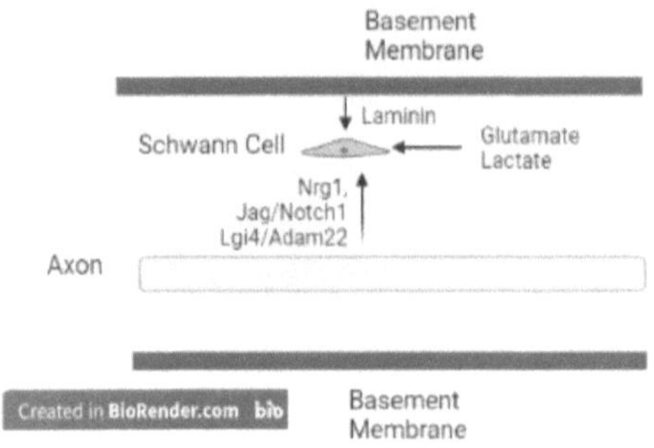

Figure 2. Microenvironmental Signals Regulating Schwann Cell Development. Schwann cells receive signals from their associated axon, the basement membrane and metabolic signals that guide maturation.

The mTOR pathway is an Integrator of Microenvironmental Signals

Upon receiving the plethora of microenvironmental inputs during development, Schwann cells must integrate these inputs to make cell fate decisions about proliferation, differentiation and myelination. Understanding how alterations in intracellular integrative signaling pathways drive Schwann cell development may be key to treating neuropathies resulting from derangements in peripheral myelination.

One signaling pathway that is known to connect growth factor and nutrient availability with cell growth is the PI3K-Akt-mTORC1 signaling pathway. The mTORC1 complex is composed of mTOR, mLST8 and Raptor[27]. Activation of mTORC1 promotes protein translation, lipid and nucleotide biosynthesis and inhibits autophagy[27]. The activity of the mTORC1 complex is increased by growth factor signaling, oxygen, energy and nutrients and decreased by stress[27,28]. Two GTPases, Rheb and Rag act to integrate growth factor and nutrient signaling for mTORC1. Binding of activated Rheb to mTORC1 drives mTORC1 signaling. However, mTOR is normally in the cytoplasm, and cannot interact with the Rheb which is normally tethered to the surface of the lysosome[29]. When Rag GTPases are activated by glucose and amino acids, they bring mTOR to the surface of the lysosome[29]. The activity of the Rag GTPases is itself regulated by the GATOR1/2 complexes downstream from cellular amino acid sensors like the sestrins and SAMTOR[30-33]. Rheb activity is also negatively regulated by the tuberous sclerosis complex, composed of Tsc1, Tsc2 and TBC1D7, which acts as a GTPase-activating protein (GAP) for Rheb[34]. Growth factor signaling through Akt, Rsk and Erk can phosphorylate Tsc2, leading the complex to fall apart and promoting activation of Rheb-mTORC1 signaling[35,36].

<u>The mTOR pathway in regulation of Schwann cell development</u>

Activation of mTORC1 signaling has been found to be required for myelination of the peripheral nervous system. Deletion of mTOR in immature Schwann cells using the *CNP-Cre* driver drives hypomyelination that does not recover with age,[37] but does not affect Schwann cell differentiation[37]. Loss of mTOR in Schwann cell precursors also leads to hypomyelination[38]. Follow up work using deletion of mTOR complex 1 adaptor molecule Raptor in Schwann cell precursors using the *Dhh-Cre* driver found that mTOR complex 1 was required for peripheral myelination[39]. Loss of Raptor in Schwann cell precursors was linked to decreased lipid and cholesterol synthesis driven by SREBP1c transcriptional activity, leading to hypomyelination[39]. In contrast, loss of mTORC1 signaling through Raptor deletion in PLP1+ mature Schwann cells had no effect on nerve myelination, but impaired Schwann cell dedifferentiation after nerve injury[40] by inhibiting metabolic reprogramming[41]. In contrast, zebrafish lacking RagA, a component of the Rag GTPase complex, show intact peripheral myelination but deficient central nervous system myelination[42]. These contrasting data may be due to either the lack of specificity in the knockout, or potentially a difference in the sensitivity of Schwann cells to nutrient dependent or growth factor dependent activation of mTOR signaling.

As loss of mTORC1 signaling was linked to hypomyelination, it was posited that overactivation of mTORC1 signaling would promote hypermyelination. Indeed, loss of PI3K phosphatase Pten in Cnp+ immature Schwann cells leads to focal hypermyelination and myelin outfoldings[43]. Pten loss in mature PLP1+ Schwann cells was also able to drive hypermyelination of peripheral nerves[44]. Genetic overexpression of constitutively active Akt using a CNP promoter increased myelin protein expression and drove focal hypermyelination and tomaculae formation in a mouse model[45]. Similarly, addition of constitutively active Akt through an adenoviral vector to rat Schwann cell nerve grafts increased myelination of the graft[46]. While these methods hint at a positive role for mTORC1 signaling in driving hypermyelination, they are not definitive as PI3K-Akt signaling has targets outside of mTOR. Loss of upstream negative mTOR regulators Tsc1 and 2 is a well validated method for activating mTOR signaling. Previous work from our lab had shown that activation of mTOR signaling by Tsc1 deletion in developing oligodendrocytes, the myelinating cells of the central nervous system, promotes hypomyelination through activation of ER stress responses[47]. We asked whether loss of Tsc1 in Schwann cells had similar effects on myelination. This question will be addressed in Chapter 1.

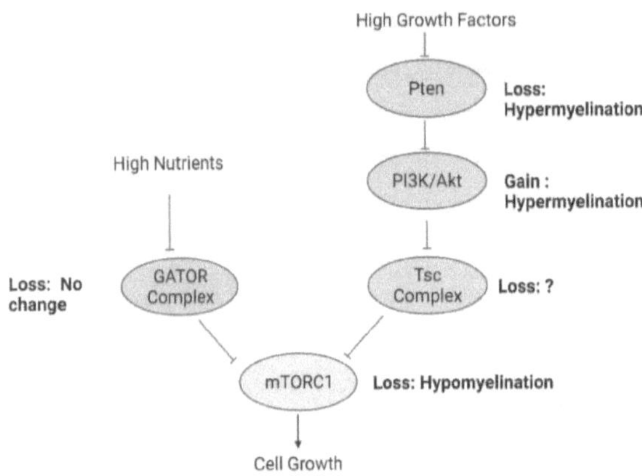

Figure 3. Summary of mTOR Signaling in Schwann Cell Myelination. Figure created using Biorender

Part II: The Microenvironment in Regulation of the Growth of Malignant Brain Tumors

<u>Genetic alterations and classification of glioblastoma</u>

Just as microenvironmental influences can regulate proliferation and differentiation of developing glia of the peripheral nervous system, so too can microenvironment signaling affect the growth of malignant brain tumors. Glioblastoma is a malignant brain tumor of the central nervous system with a median survival of 14.6 months. Current management for glioblastoma includes combinations of surgery, radiation and chemotherapy. Adult glioblastomas are characterized by activation of RTK signaling by amplifications of EGFR and PDGFRA, activation of MAPK signaling through loss of NF1, activation of PI3K signaling through loss of PTEN and activating mutations in PI3-kinase

(PIK3CA, PIK3R1), epigenetic derangements through mutations in IDH1, loss of DNA repair regulation through loss of TP53 and activation of cell cycle signaling through inactivating mutations in RB and CDKN2A[48,49]. However, targeted inhibition of these tumor intrinsic signaling pathways has not yet proven effective in treating patients[50-52].

Transcriptional profiling of glioblastoma has identified two major subclasses of tumors based on mutations of the IDH1/2 enzymes. Mutation of IDH enzymes leads to loss of its normal function of converting isocitrate to alpha-ketoglutarate, and a gain of function converting alpha-ketoglutarate to the new oncometabolite 2-hydroxyglutarate (2-HG). This new metabolite inhibits the function of Tet DNA demethylases and histone demethylases leading to epigenetic reprogramming of tumor cells[53]. IDH mutant tumors, also called G-CIMP tumors due to their DNA hypermethylation, tend to have a better prognosis than other glioblastomas[54,55]. Methylation profiling further subdivides IDH mutant tumors into three subgroups: one with 1p/19q co-deletion with good prognosis, one with high levels of methylation (G-CIMP high) with good prognosis, and one with low levels of methylation (G-CIMP low) with poor prognosis[56].

IDH WT tumors can be grouped into three separate subtypes: classical, proneural and mesenchymal which have different drivers and sensitivity to treatment[57]. Classical tumors are characterized by amplifications of EGFR, EGFRvIII truncations, loss of PTEN, no TP53 mutations and loss of CDKN2A[58]. Classical tumors are enriched in astrocytic gene signatures[58], and are most likely to benefit from aggressive chemotherapy regimens[48,58]. The proneural subtype of glioblastoma has enrichment of PDGFRA amplification (35%) and mutations in TP53 (54%)[58,59]. Proneural subtype tumors also show enrichment of genes associated with oligodendrocyte development[58] and have been linked to SOX10 transcriptional activity[60]. Addition of anti-angiogenic therapy to standard chemotherapy regimens has been shown to improve survival in the proneural subtype[61,62]. The mesenchymal subtype of glioblastoma is enriched in NF1 deletions (38%) and TP53 mutations (32%)[58]. Mesenchymal gene signatures have been associated with quiescence in glioma stem cell populations[63]. Mesenchymal differentiation is also enriched in recurrent tumors, and is associated with radioresistance and resistance to anti-angiogenic therapy[64-67]. Mesenchymal differentiation is associated with increased canonical and non-canonical Nf-κB signaling in tumor cells[66,68], and master regulator analysis identifies Stat3, CEBP/B and TAZ as drivers of the mesenchymal glioblastoma[69,70]. The mesenchymal subtype of glioblastoma has also been long recognized as having a greater degree of immune infiltration and activation, with a higher percentage of tumor-associated macrophages when compared to other subtypes[57,71], and a greater level of angiogenesis when compared to other subtypes[72].

In addition to transcriptional heterogeneity between different tumors, glioblastomas also display a significant amount of intratumoral heterogeneity. Individual tumors have been found to express multiple receptor tyrosine kinases and to possess clones from multiple transcriptional subtypes within glioblastoma[73-76]. Single cell profiling has also identified that single cells in glioblastoma exist on a continuum of differentiation between OPC-like, neural progenitor-like, astrocyte-like and mesenchymal-like states driven by different genetic events[77]. Another group, using single cell RNA-seq and single nucleus ATAC-seq, found that proliferating glioma stem cells in human tumors exist on a continuum from proneural GSCs to mesenchymal GSCs[78]. Analysis of clonal dynamics in glioblastoma demonstrates the co-existence of multiple independent subclones over Darwinian selection from primary human tumors, even after serial passaging in mice[79]. Heterogeneity in tumor transcriptional state may be driven by the local milieu of the tumor cells; one group found the tumor core is enriched for mesenchymal GSCs in a hypoxic niche, while the infiltrating tumor edge is enriched for proneural GSCs[80]. The heterogeneity of tumor cells in glioblastoma makes it a difficult target for therapy.

<u>The tumor microenvironment is a novel target for anti-tumor therapy in glioblastoma</u>

The tumor microenvironment is the non-malignant component of a cancer composed of infiltrating and resident immune cells, resident stromal cells, and abiotic factors such as hypoxia and metabolites. As microenvironmental cells are more genetically stable than the constantly mutating tumor population, they represent an attractive alternative for anti-cancer therapy. Targeting components of the tumor microenvironment through activation of lymphocytes, inhibition of angiogenesis or blockade of tumor associated macrophage function has led to dramatic responses on tumor growth and significant extension of overall survival in many cancers[81-83].

The glioblastoma microenvironment is composed of astrocytes, oligodendrocytes, neurons, lymphocytes, granulocytes, vascular cells, and microglia/macrophages. Emerging research has begun to elucidate the function of these cells in growth and treatment response. Tumor-associated astrocytes have been linked to immunosuppression[84], promotion of tumor invasion[85], and drug resistance through nanotube networks[86]. In a GL261 mouse glioma transplant model, perivascular tumor-associated oligodendrocyte progenitor cells (OPCs) promote vascular sprouting at the invasive front of glioma[87]. Excitatory glutamatergic neuronal activity can act to stimulate the growth of gliomas, while inhibitory neuronal activity inhibits glioma proliferation[88-92]. Tumor-infiltrating lymphocytes are a minor population in human glioblastomas, with a low level of T cell infiltration and high myeloid infiltration, corresponding to a "immune desert" subtype[93]. . As a result, immune checkpoint blockade targeting PD-1 or CTLA-4 is not efficacious in most patients, except for subsets with MAPK pathway activation[94-96]. Infiltration of granulocytes in glioblastoma has been linked to ferroptosis[97], response to anti-angiogenic therapy combined with chemotherapy[98], stimulation of glioma proliferation and resistance to anti-VEGF therapy[99], and resistance to anti-PD-1 blockade[100,101].

An actively growing vasculature is a requirement for tumor to obtain enough nutrients to grow well. The brain vasculature is unique in that it forms a blood-brain barrier composed of endothelial tight junctions, pericyte coverage and astrocytic endfeet[102]. In malignant brain tumors, high levels of angiogenic factors promote the formation of a leaky, tortuous vasculature that has partial blood brain barrier breakdown[103]. However, the vasculature also expresses high levels of drug efflux transporters, limiting drug penetration into malignant brain tumors[104]. Growth of the tumor vasculature also sets up different niches based on hypoxic vs perivascular areas that support the maintenance of glioma stem cells[105]. In glioma, the VEGFA-VEGFR2 pathway acts to stimulate neovessel sprouting. However, blockade of this pathway has only demonstrated modest improvements in survival in clinical trials, suggesting the need for investigation into alternative angiogenesis pathways or rational combination therapies[61,62,106-109]   .

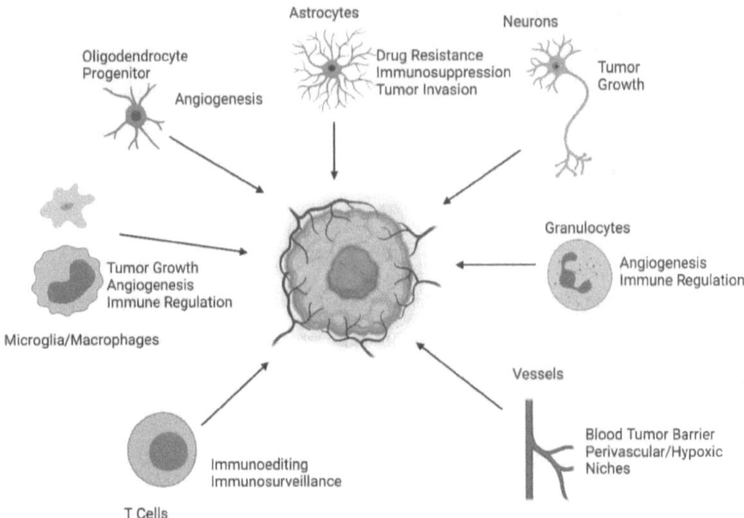

Figure 4. Summary of the Tumor Microenvironment in Glioblastoma. Figure created using Biorender

<u>Macrophages in the CNS originate from three lineages and have differential contributions to brain tumorigenesis</u>

The most prominent microenvironment cell type in human glioblastoma are tumor-associated macrophages. Macrophages in the CNS can be divided into microglia, CNS border associated macrophages and monocyte-derived macrophages[110]. Microglia are derived from yolk sac-derived macrophages that migrate into the brain during embryonic development[111]. They play important roles in synaptic pruning[112] and local neuroinflammation, but are not very good at presenting antigens to infiltrating T cells[113]. A second population of CNS macrophages are the border-associated macrophages. This population of macrophages is present at CNS interfaces such as the perivascular space, perimeningeal space and choroid plexus[113]. They have been reported to play roles in antigen presentation and scavenging, but their role in both normal and disease states is incompletely understood[114]. A third population of macrophages that can enter the CNS in pathological states are monocyte-derived macrophages. During inflammation, blood derived monocytes can enter the CNS and differentiate into microglia-like cells. Antigen presentation by monocyte-derived macrophages rather than CX3CR1+ microglia or border macrophages is required for the development of EAE in mice[113]. This Cd49d+ population localizes to the tumor core of mouse and human gliomas, expresses molecules that support angiogenesis and tumor growth, and is associated with poor prognosis in human gliomas[115-117]. Monocyte-derived macrophages also are thought to promote radiation survival in the post-treatment setting[118]. Other reports have suggested that tumor-associated microglia rather than macrophages are the most relevant pro-tumor population that promote angiogenesis[119].

<u>Tumor-associated macrophages regulate glioma growth</u>

Most existing work in the field speaks to a pro-tumor role of glioma-associated macrophages. Tumor associated macrophages were first defined using the M1-M2 paradigm. M1 macrophages, driven by Nf-κB signaling induced by TLR ligation, IFNγ stimulation or TNF signaling, express pro-inflammatory genes and are thought to be anti-tumor. M2 macrophages correspond to IL-4/IL-13 stimulation and express wound repair, immunosuppressive and angiogenic molecules[120]. Early characterization of glioma associated macrophages defined them as having an M2-like phenotype based on expression of classic M2 marker genes such as *Mrc1* and *Cd163*. Manipulation of these M2-like cells was able to affect tumor growth. Intratumoral depletion of M2-like macrophages using a MAFIA mouse model decreased tumor growth[121]. Conversely, cotransplant of M2-polarized U937 macrophage-like cells with glioma stem cells has also been shown to accelerate tumor growth[122], through secretion of pleiotrophin. Reductions in tumor growth were also observed in studies that used a head-specific *Cd11b-TK* system, in which the head was protected from irradiation[119]. However, as more profiling of tumor-associated macrophages has been performed, both human and mouse TAMs have been defined as a transcriptionally distinct population from M1/M2 macrophages[115,123,124]. While studies have identified specific functions for glioma TAMs in regulation of growth[125,126], and immunosuppression[127], the function of these cells in heterogeneous human glioblastomas remains poorly defined.

Glioma-associated macrophages are an important regulator of blood vessel growth. In tumors resistant to bevacizumab, there is increased recruitment of an angiogenic population of macrophages expressing Tie2 and Nrp1 that promote relapse[128-130]. In other contexts, macrophage stimulation of vessel growth can inhibit tumor growth. In a GL261 model where macrophages were depleted via anti-CSF1R antibody treatment, tumor growth accelerated following macrophage depletion due to normalization of the vasculature and increased perfusion of the tumor[103]. Normalization of the vasculature also improves delivery of chemotherapies to the brain tumor microenvironment, as well as promotes the infiltration of anti-tumor immune cells[103,131]. Normalization of vasculature with combination of Ang2 blockade and VEGFR2 inhibition delayed glioma growth in a macrophage-dependent manner in GL261 model of glioma[132].

<u>Reprogramming of tumor associated macrophages can promote tumor regression</u>

Targeting of tumor associated macrophages has been explored through small molecular inhibitors of the CSF1-CSF1R pathway, which is involved in recruitment, proliferation and polarization of tumor-associated macrophages[133]. Treatment of mouse Pdgfb-driven gliomas with CSF1R inhibitor BLZ945 has been shown to promote durable regression and dramatically improved long-term survival of tumor-bearing mice[134]. Tumors that became resistant to BLZ945 showed activation of PI3K signaling in tumor cells, and loss of Pten blunted responses to this drug[135]. Targeting of the CSF1R signaling pathway also has been linked to increased response to tyrosine kinase inhibitors in a Pdgfb-driven model of glioma[136]. However, a phase II clinical trial of CSF1R inhibitor PLX3397 in human glioblastomas only showed a treatment response in two out of twelve patients, suggesting perhaps intertumoral heterogeneity may regulate response to tumor macrophage targeting[137]. Other signaling regulators such as PI3Kγ signaling[138] and type II HDACs[139] have been described as methods to target macrophage polarization in cancer. We asked the question whether response to CSF1R inhibition was dependent on tumor subtypes in glioblastoma. This question will be addressed in Chapter 2.

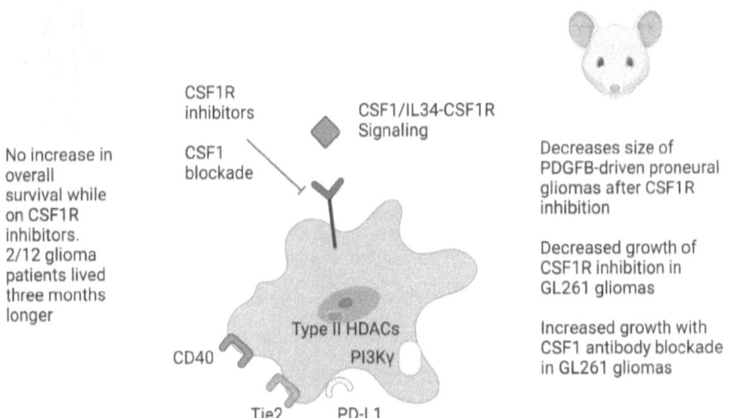

Figure 5. Targeting of Tumor Associated Macrophages. Summary of targets of tumor associated macrophages in cancer. Figure created using Biorender

Part III: Remodeling of the Tumor Microenvironment by Radiation Therapy in Glioblastoma

Radiation therapy is an important pillar of current glioblastoma treatment regimens. Radiotherapy promotes DNA double strand breaks and membrane damage through the creation of free radicals. In addition to tumor intrinsic effects, radiation therapy induces a systemic immune response which is capable of inducing tumor control even at distant sites that were not treated through the abscopal effect[140]. The mutations induced by irradiation also promote the generation of neoantigens that serves as targets for endogenous T cell responses[141]. Indeed, in some preclinical models, addition of immune checkpoint blockade to radiotherapy has shown synergistic response in vivo[142]. In glioblastoma, response to radiation therapy is usually transient; the tumors eventually recur and end up killing the patient. While patients with recurrent gliomas are often the subjects for clinical trials, surprisingly little is known about the biology of recurrent glioblastoma when compared to control tumors.

Treatment of glioblastomas induces remodeling of tumor cell populations. One model of understanding the biology of recurrent glioblastomas is the cancer stem cell hypothesis. This hypothesis argues that just as normal tissues are maintained by a population of slowly dividing stem cells, malignant tumors also possess a population of slowly dividing tumor cells that have the capacity to give rise to the entire tumor. In treatment naïve tumors, these cells usually compose of a minority of cells in the tumor, but when chemoradiation depletes rapidly dividing differentiated cells, cancer stem cells will become activated and proliferate to form the recurrent tumor. Glioma stem cells show increased activation of DNA damage response and drug efflux transporters, making them resistant to radiation therapy and chemotherapy[143,144]. In a hGFAP-Cre Nf1$^{f/+}$ p53$^{f/f}$ Pten$^{f/+}$ genetically engineered mouse model, chemical debulking of proliferating cells using temozolomide demonstrated that recurrent tumors arose from a Nestin+ cancer stem cell[145]. Adult primary and recurrent glioblastomas tend to share similar transcriptional profiles in patients treated with conventional radiotherapy/temozolomide[146], and about 55% of human tumors in a recent dataset maintained the same transcriptional subtype following recurrence[57]. Of the tumors that switched subtype, proneural and mesenchymal subtypes were increased after recurrence, while classical subtype tumors were decreased[57]. Recurrent tumors also show increased FGFR1 signaling, which is associated with poor survival outcomes[147].

The microenvironment is also altered in recurrent tumors. Tumor cells implanted in an irradiated microenvironment grow much faster than those implanted in a non-irradiated microenvironment[148].

This effect is partially mediated by tumor associated- astrocytes which secrete Tgm2 after radiaton therapy to promote stemness and radioresistance of tumor cells[149]. Recurrent tumors showed higher infiltration of M2 tumor-associated macrophages and immature dendritic cells when compared to primary tumors[57,150]. Targeting of tumor-associated macrophages significantly extends survival after radiotherapy[118,151] and improves local control of glioblastoma in human patients[152]. Radiotherapy induced senescence of tumor cells promotes recruitment of granulocytic cells, which act to promote the conversion of differentiated glioma cells to glioma stem cells[153]. Radiotherapy has been associated with an increase in T cell infiltration in recurrent tumors after radiotherapy[154], and increased regulatory T cell infiltration, promoting immunosuppression[155]. Recurrent tumors also show a decrease in microvessel density, an increase in vessel diameter, and decreased water diffusion when compared to parent tumors[156]. The recurrent tumor microenvironment is therefore an important regulator of glioma relapse and immune escape.

Conventional photon radiation therapy often leads to long term neuro-cognitive deficits due to the effect of radiation on the developing brain[157]. Proton therapy is a method that has garnered an interest as a potential alternative to conventional photon radiotherapy. Ionized protons deliver the same biologically effective dose as conventional x-rays but have additional properties that enhance tissue sparing. Namely, proton radiation shows a sharper drop off in radiation dose deposited after a certain tissue depth, known as the Bragg peak, when compared to conventional x-rays, allowing for increased tissue sparing. In pediatric patients, those treated with proton irradiation had better cognitive outcomes when compared to those treated with x-rays[158] and proton therapy has been found to reduce grade 3+ lymphopenia in adult patients[159]. Patients treated with proton therapy also did not see the decreases in grey and white matter volume when compared to photon treated patients[160]. Proton therapy has been deployed to treat glioblastomas both therapeutically[161,162], and palliatively[163]. However, the changes in both the tumor cells and tumor microenvironment induced by proton therapy remain poorly understood, presenting a barrier to targeted treatment of recurrent gliomas following proton irradiation. We sought to answer this question by treating mice from an endogenous model of glioma with whole brain proton irradiation and then performing single cell RNA-seq to characterize changes in the tumor and microenvironment after proton therapy. This data will be discussed in Chapter 3.

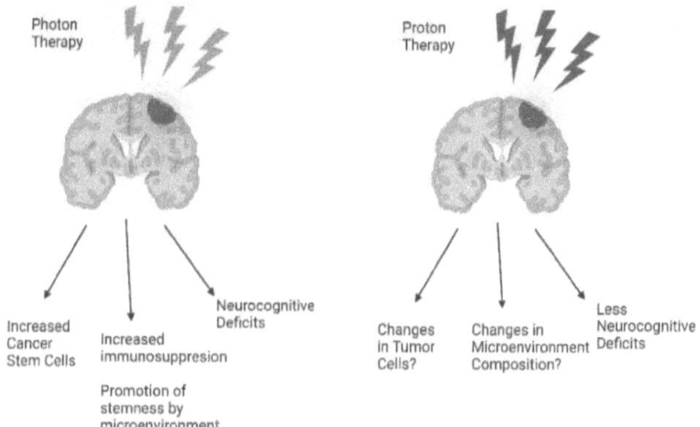

Figure 6. Proton and Conventional Photon Treatment Effects on Recurrent Tumors and Brain Function

Chapter 1: The TSC1-mTOR-PLK Axis Regulates the Homeostatic Switch from Schwann Cell Proliferation to Myelination in a Stage-Specific Manner

Introduction

Myelination of axons is vital for rapid saltatory conduction of action potentials and effective sensorimotor integration. In the peripheral nervous system (PNS), Schwann cells (SCs) wrap large-caliber axons to form tightly compacted myelin sheaths. Defects in myelination lead to peripheral neuropathies such as Charcot-Marie-Tooth disease and congenital hypomyelinating neuropathy[1-4].

Peripheral myelin formation requires SC expansion and differentiation, and the onset of myelination is linked to cell cycle exit[164,165]. Controlling the balance between SC proliferation and differentiation is critical for ensuring myelination proceeds properly. The TSC1/2-mTOR signaling axis regulates cell growth and differentiation by regulating protein translation[166]. Various upstream signals, such as nutrients and growth factors, converge on the tumor suppressors TSC1 and TSC2, which form a complex that negatively regulates the mTOR pathway, and ultimately organismal growth and homeostasis[28,167]. Disruption of the TSC1/2 complex activates Rheb-GTPase and subsequent mTOR signaling through the mTORC1 and mTORC2 complexes. mTOR targets S6K and 4EBP1 control cell growth processes such as ribosomal biogenesis and cap-dependent protein translation[168,169].

SC growth and differentiation can be driven by extracellular molecules such as neuregulin I (NRG1) and laminin 211/GPR126, and intracellular signaling cascades including the Hippo and PI3K-AKT signaling pathways[11,16,170-173]. Activation of the PI3K-AKT pathway by axonal NRG1 acts upstream of mTOR, regulating SC migration, proliferation, and survival[174-176]. Balanced effects of cell-cycle promoting effectors and inhibitors, such as CDK2 and p27, respectively, regulate SC lineage progression[177,178]. In addition, mTOR signaling has also been implicated in regulation of the polo-like kinase (PLK) pathway[179,180]. PLKs are key regulators of cell cycle progression with established functions in cell cycle entry, mitosis, bipolar spindle formation, chromosome segregation, and cytokinesis[181-183].

mTORC1 is crucial for myelination in both central and peripheral nervous systems[37,39,184-188]. Loss of mTOR leads to profound hypomyelination in the PNS[37], and, conversely, activation of mTOR through the loss of PTEN or activation of AKT promotes hypermyelination in the PNS[44,45,175]. Recent studies indicate that TSC1/2-mTOR signaling is a critical regulator of SC development and myelination in the PNS[184,189], however, the precise mechanisms whereby mTOR controls SC growth and the transition from proliferation to differentiation are not yet fully understood

Preliminary Data

To test the effect of hyperactive mTOR in Schwann cell myelination, my co-first author knocked out upstream mTORC1 inhibitor Tsc1 using a Dhh-Cre promoter, which is first expressed in Schwann cell precursors at E12.5 (Tsc1 cKO mice) (Figure S1A). Sciatic nerves collected at P10 demonstrated increased mTORC1 activity, and a decrease in myelin protein expression (Figure S1B-D). Mice with conditional heterozygous knockout of Tsc1 demonstrated no change in myelin thickness (Figure S2).

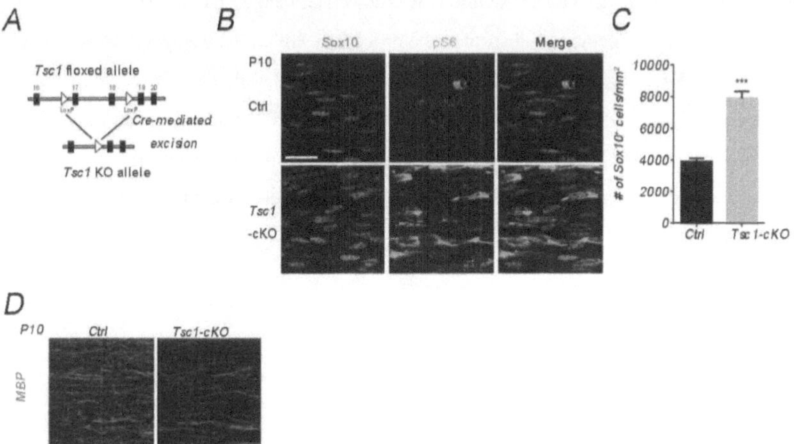

Figure S1. Schwann-cell specific loss of Tsc1 hyperactivates mTOR and drives hypomyelination in developing sciatic nerves. A) Diagram of strategy for deletion of exons 17 and 18 of Tsc1 by Dhh-Cre expressed specifically at E12.5 in Schwann cells, B) Representative images of Sox10 and p-S6 staining of control and Tsc1-cKO sciatic nerves at P10. Scale bar, 15 µm, C) Quantification of Sox10+ Schwann cells per mm2 (n = 6 animals/genotype). Student's t-test, *** p<0.001. D) Representative images of control and Tsc1-cKO sciatic nerves at P10 stained for MBP. Scale bar, 25 µm.

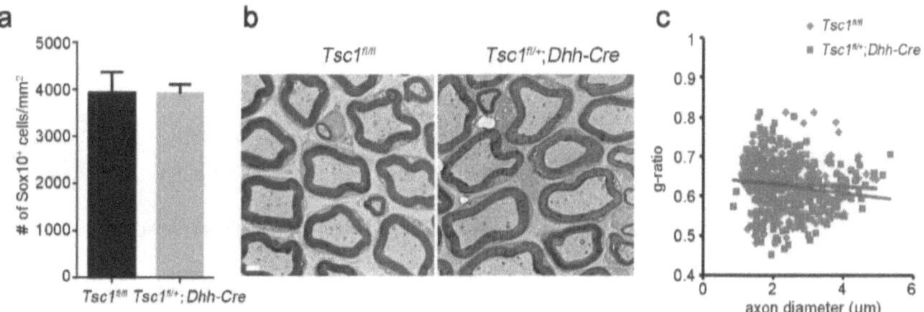

Figure S2. SC number and myelination in Tsc1fl/fl and Tsc1fl/+;Dhh-Cre peripheral nerves are comparable:(a) Quantification of the proportions of Sox10+ cells in Tsc1fl/fl and Tsc1fl/fl;DhhCre sciatic nerves at P30 (n = 6 animals/genotype). Student's t-test, (b) Electron micrographs showing the ultrastructure of sciatic nerves from Tsc1fl/fl and Tsc1fl/fl;DhhCre mice at P30. Scale bar, 2 µm, (c) Quantification of g-ratios of P30 Tsc1fl/fl and Tsc1fl/fl;DhhCre sciatic nerves. n = 3 animals/genotype. Student's t-test.

Electron micrographs of the myelin ultrastructure at P7 showed an increased number of
unmyelinated axons, and a defect in radial sorting, the process by which Schwann cell associate 1:1
with axons (Figure S3A-C). Examination of Tsc1 cKO sciatic nerves at P30 and P60 demonstrated
a persistently increased number of unmyelinated axons when compared to matched controls and a
lower myelin thickness (Figure S3D-I). Four-month-old Tsc1 cKO mice also showed the
appearance of axons with redundant myelin formation that extended out from the myelin membrane
or compressed the axon (Figure S3J-L).

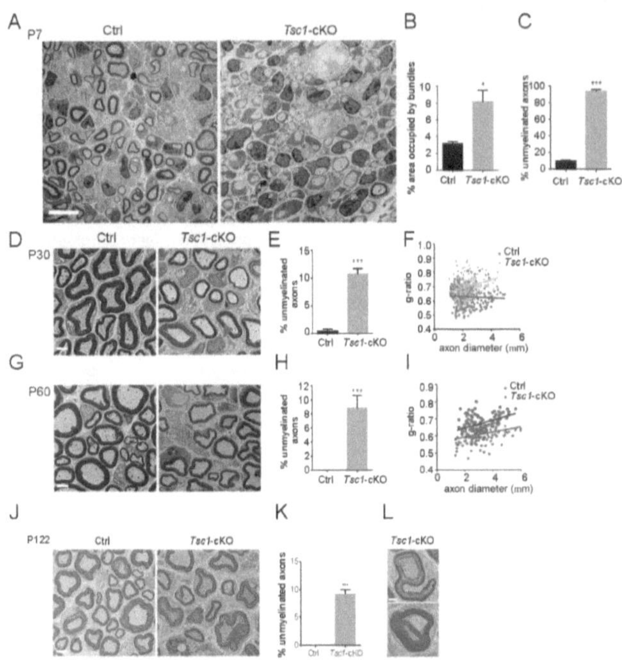

Figure S3. Schwann Cell-Specific Loss of Tsc1 Delays Radial Sorting and Promotes Dysmyelination. A)
Representative electron micrographs showing ultrastructure of sciatic nerves in control and Tsc1-cKO at P7.
Unmyelinated large axon bundles are indicated with an arrowhead. Scale bar, 10 μm, B) Quantification of the areas of
large bundles in control and Tsc1-cKO sciatic nerves at P7. n = 3 animals/genotype. Student's t-test, * p<0.05, C) The
percentages of unmyelinated axons in P7 control and Tsc1-cKO sciatic nerves. n = 3 animals/genotype. Student's t-test,
*** p<0.001, D) Electron micrographs showing the ultrastructure of sciatic nerves from control and Tsc1-cKO mice at
P30. Arrowheads indicate SCs that are not associated with axons. Scale bar, 2 μm, E) Quantification of percentages of
unmyelinated axons in P30 control and Tsc1-cKO sciatic nerves. n = 3 animals/genotype. Student's t-test, *** p<0.001.
F) Quantification of g-ratios of P30 control and Tsc1-cKO sciatic nerves. n = 3 animals/genotype. Student's t-test, ***

p<0.001, G) Electron micrographs showing ultrastructure of sciatic nerves from control and Tsc1-cKO mice P60. Scale bar, 2 μm, H) Quantification of percentages of unmyelinated axons in P60 control and Tsc1-cKO sciatic nerves. n = 3 animals/genotype. Student's t-test, *** p<0.001, I) Quantification of g-ratios of P60 control and Tsc1-cKO sciatic nerves. n = 3 animals/genotype. Student's t-test, *** p<0.001. J) Representative electron micrographs of P122 control Tsc1fl/fl and Tsc1-cKO sciatic nerves. Scale bar, 2 μm, K) The percentages of unmyelinated axons in P122 control and Tsc1-cKO sciatic nerves. n = 3 animals/genotype. Student's t-test, *** p<0.001, L) High magnification electron micrographs showing redundant myelin in P122 Tsc1-cKO sciatic nerves. Scale bar, 800 nm.

Differentiation of Schwann cells from immature Schwann cells to myelinating Schwann cells proceeds postnatally through defined phases. Immature Schwann cells expressing the marker Sox2 proliferate, undergo radial sorting and upregulate Oct6 to differentiate into pro-myelin Schwann cells. Pro-myelin Schwann cells then downregulate expression of Oct6 and Sox2, exit the cell cycle and differentiate into Krox20+ myelinating Schwann cells. The combination of Sox10+ Schwann cell lineage expansion with decreased myelin protein expression suggested a block in differentiation. Tsc1 cKO sciatic nerves demonstrated an increase in pro-myelin Oct6+ and immature Sox2+ Schwann cells, and a decrease in mature Krox20+ Schwann cells (Figure S4A, B, G, H). Tsc1 cKO sciatic nerves also show an increase in proliferating cells in the Sox10+ Schwann cell lineage (Figure S4C-F). These data are consistent with a model where Tsc1 loss arrests developing Schwann cells in the proliferative, pro-myelin stage.

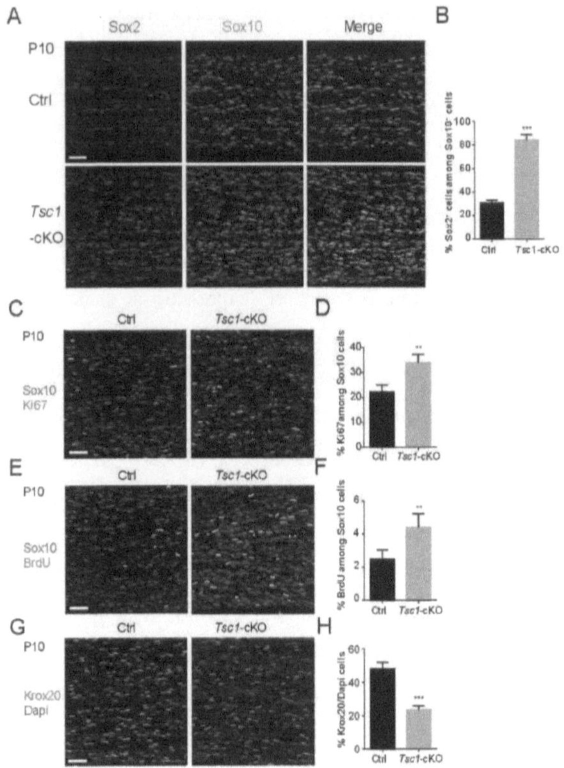

Figure S4. Schwann cell knockout of Tsc1 show an arrest in differentiation at the pro-myelin stage A) Representative images showing Sox2 and Sox10 staining of control and Tsc1-cKO mutants at P10. Scale bar, 50 μm, B) Quantification of percentages of Sox2+ cells among Sox10+ Schwann cells (n = 6 animals/genotype). Student's t-test, *** p<0.001. C) Representative images control and Tsc1-cKO sciatic nerves at P10 immunostained for Sox10 and Ki67. Scale bar, 50 μm, D) Quantification of cells expressing both Ki67 and Sox10 in control and Tsc1-cKO sciatic nerves (n = 6 animals/genotype). Student's t-test, **p<0.01, E) Representative images P10 control and Tsc1-cKO sciatic nerves stained for Sox10 and labeled with BrdU. Scale bar, 50 μm, F) Quantification of cells expressing both BrdU and Sox10 in control and Tsc1-cKO sciatic nerves (n = 6 animals/genotype). Student's t-test, **p<0.01, G) Representative images showing Krox20 expression in control and Tsc1-cKO sciatic nerves at P10. Scale bar, 50 μm, H) Percentages of Krox20+ cells in P10 control and Tsc1-cKO sciatic nerves (n = 6 animals/genotype). Student's t-test, *** p<0.001.

To determine if downstream mTOR signaling mediated the Tsc1-induced hypomyelination, my co-first author crossed mTOR flox mice with Dhh-Cre mice to generate Dhh-Cre Tsc1 f/f mTOR f/+

and Dhh-Cre Tsc1 f/f mTOR f/f mice. Ultrastructural analysis of these mice revealed that loss of one allele of mTOR was sufficient to increase myelin thickness in Tsc1 cKO sciatic nerves (Figure S5). Loss of both alleles of mTOR led to hypomyelination, consistent with the literature finding with mTOR loss alone.

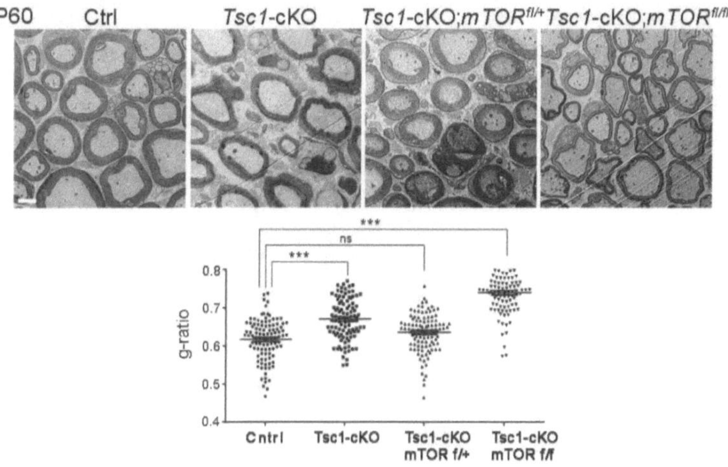

Figure S5. Loss of one allele of mTOR rescues myelination defects in Tsc1 cKo mice. A) Representative electron micrographs of ultrastructure from P60 control, Tsc1-cKO, Tsc1-cKO;mTORfl/+ and Tsc1-cKO;mTORfl/fl. Scale bar, 2 μm, B) Quantification of g-ratios in P60 control, Tsc1-cKO Tsc1-cKO;mTORfl/+, Tsc1-cKO;mTORfl/fl sciatic nerves (n > 200 myelinated axons from 3 mice per group). One way ANOVA with Tukey's t-test used to determine significance.

Finally, to elucidate the pathways downstream of mTOR mediating hypomyelination, RNA-seq was performed on P7 control and Tsc1 cKO sciatic nerves. Consistent with the gross hypomyelination, Tsc1 cKO nerves showed depletion of genes associated with myelination, PI3K-Akt signaling, and lipid biosynthesis (Figure S6A,B). Concomitantly, mutant nerves showed enrichment of genes associated with cell cycle proliferation, Polo-like kinase pathway, and negative regulation of myelination (Figure S6A,B). The Polo-like kinase pathway was chosen for further analysis because of its role in driving cell cycle progression, and potential druggability. Validation of RNA-seq results via qPCR showed upregulation of Polo-like kinase pathway genes Plk1, Plk2, Cdc25a and Cdc25b (Figure S6C). Plk1 expression was found to be upregulated via Western blot (Figure S6D).

Similar to responses to Tsc1 loss in the central nervous system, there was an upregulation of ER stress response in Tsc1-cKO Schwann cells (Figure S7).

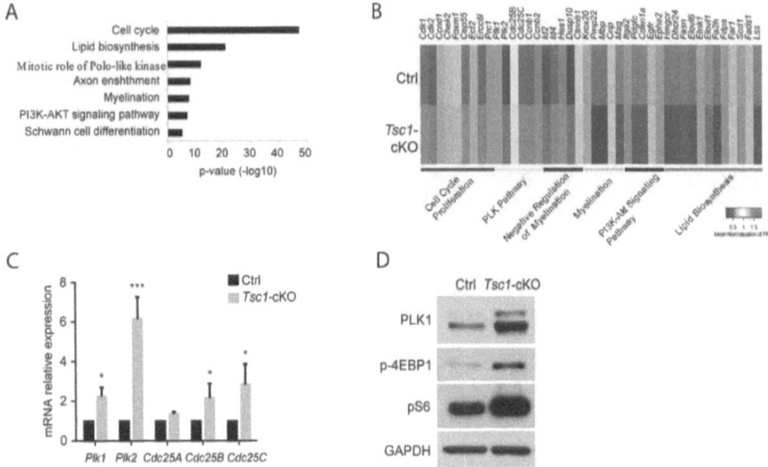

Figure S6. Tsc1-cKO Sciatic Nerves are Enriched in Genes Associated with Polo-like Kinase Signaling. A) Gene ontology analysis of pathways altered in P7 Tsc1-cKO sciatic nerves compared with their control littermates, B) Heat map of differentially regulated pathway in Tsc1-cKO sciatic nerves compared with control littermates, C) qRT-PCR of mRNA expression of PLK pathway genes in control and Tsc1-cKO sciatic nerves at P7. n = 3 independent experiments. Student's t-test, ***p<0.001, *p<0.05, D) Western blot showing expression of PLK1, p-S6, and p-4EBP1 in P7 control and Tsc1-cKO sciatic nerves. Blots are representative of triplicate experiments. GAPDH was used as a loading control.

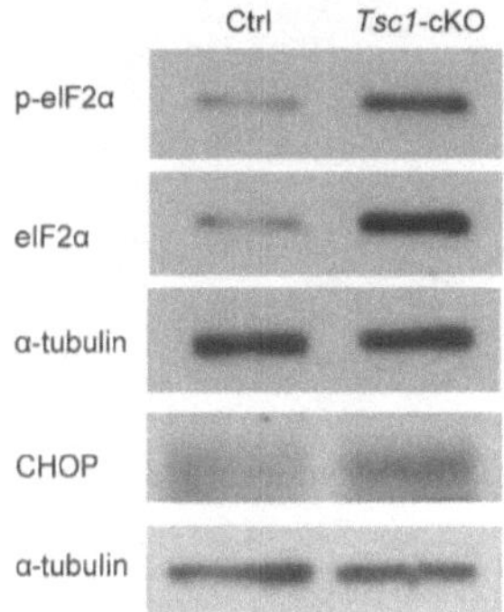

Supplementary Figure S7. *Tsc1* **ablation elevates ER stress responses in SCs:**(a) Representative western blot of expression of ER stress response proteins, p-eIF2α, eIF2α, and CHOP in *Tsc1*-cKO mutants and control sciatic nerves. α-tubulin was loaded as a control.

Results

Dhh-Cre Mediated Excision of Tsc1 Decreases Expression of Tsc1 in Sciatic Nerves

The preliminary data had shown that Tsc1 cKO nerves had higher levels of mTORC1 signaling, as evidenced by phospho-S6 (Figure S1B). However, there was no direct evidence showing Tsc1 loss. To explore this question, I attempted Western blots of Tsc1 on control and knock out nerve lysates. Unfortunately, several attempts to assess protein expression via Western blot failed, likely due to the large molecular weight of the protein. I next decided to compare control and mutant sciatic nerves via immunofluorescence. However, as whole nerve staining would likely not provide images of sufficient quality due to the ubiquitous expression of Tsc1 in control nerves, I learned how to generate teased fiber samples in order to image individual nerve fibers. I then took teased fiber samples from control and Tsc1 mutant nerves and then performed Tsc1 immunofluorescent staining. Consistent with our phospho-S6 staining results, I found loss of Tsc1 immunofluorescence in Tsc1 conditional knockout sciatic nerves (Figure 1B). To verify that loss of Tsc1 in developing Schwann cells leads to hypomyelination, I performed a Western blot of myelin basic protein (MBP), a marker of mature myelinating Schwann cells. . In mice, there are ten classic isoforms of MBP ranging from 13 kDa to 21.5 kDa[190]. I performed this western blot and saw a decrease in protein expression of MBP with multiple bands corresponding to the major isoforms, consistent with hypomyelination in Tsc1-cKO nerves. (Figure 1C).

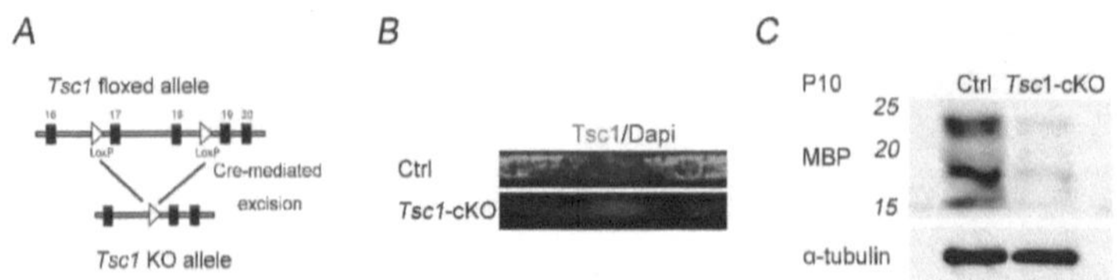

Figure 1. **Dhh-Cre Mediated Excision of Tsc1 in Developing Schwann Cells Impairs Myelination.** A) Schematic showing deletion in Tsc2-binding domain of Tsc1 by Cre recombination, B) Teased fiber staining of Tsc1 in control and Tsc1 conditional knockout sciatic nerves, C) Western blot of myelin basic protein in both control and Tsc1 cKO sciatic nerves

Loss of Tsc1 in Mature Schwann Cells Promotes Formation of Redundant Myelin

Our preliminary data shows that when followed to adulthood, Dhh-Cre Tsc1 cKo mice would develop redundant myelin. Because this effect was the opposite of what we had seen in early development, I hypothesized that increased myelin growth was a result of Tsc1 loss and mTOR hyperactivity in mature Schwann cells. To isolate this effect, I crossed Plp-CreERT mice with Tsc1 flox/flox mice to generate conditional knockout mice (Tsc1 iKO) that would ablate Tsc1 in mature Schwann cells following tamoxifen administration. I administered two five-day pulses of tamoxifen starting at P30 and then followed the mice for the next three months (Figure 2A). At four months (P120), I isolated the nerves for electron microscopy and took images of the ultrastructure. I found that similar to the long term follow up Dhh-Cre Tsc1 f/f mice, conditional knockout of Tsc1 in mature Schwann cells led to formation of redundant myelin (Figure 2B). Myelination of axons depends on axon diameter; large axons have thick myelin sheaths while axons with the smallest diameter are largely unmyelinated. We found the g-ratio for small and medium diameter axons was decreased, consistent with increased myelin thickness (Figure 2C). I then asked the question whether mTOR hyperactivity in mature Schwann cells induced by Tsc1 induced similar changes in proliferation and differentiation when compared to developmental Tsc1 knockout. I performed a small pilot experiment to look at the early changes in sciatic nerves 44 days after the start of tamoxifen induction. Consistent with increased mTORC1 signaling, I found increased levels of p-S6 in Tsc1 inducible knockout (iKO) nerves when compared to control nerves (Figure 2D). In contrast to Tsc1 cKO mice, proliferating Ki67+ cells were very rare (Figure 2D). Consistent with this, the number of Sox10+ cells was not significantly increased in Tsc1 iKO sciatic nerves (Figure

2D). Finally, in contrast to control sciatic nerves in which there were no Oct6+ cells, Tsc1 iKO
nerves showed numerous cells expressing this transcription factor from early development (Figure
2D).

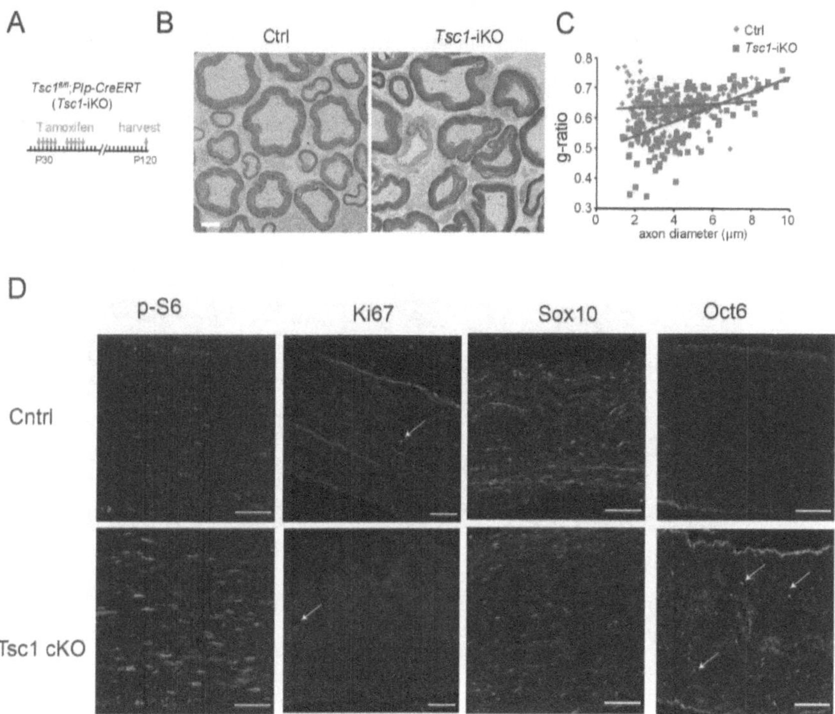

**Figure 2. Hyperactivation of mTOR Signaling in Mature Schwann Cells Promotes Redundant Myelin
Formation.** A) Schematic for knockout of Tsc1 in mature Schwann cells using the PLP-CreERT driver, B)
Representative electron micrographs of Tsc1 f/f control and Tsc1-iKO mice at P120 (N=3 mice per group). Scale bar, 2
uM, C) Quantification of g-ratio in control and Tsc1-iKO sciatic nerves (n > 200 axons from 3 mice per group).
Student's t-test, ***p<0.001, D) Representative staining of P74 control and Dhh-Cre Tsc1 f/f sciatic nerves for p-S6,
Ki67, Sox10, and Oct6 (n=2 animals per group). Scale bar, 100 uM

Loss of one allele of mTOR rescues differentiation defects associated with mTOR
hyperactivity

The preliminary data had shown that loss of one allele of mTOR showed a rescue of
hypomyelination in Tsc1 conditional knockout nerves. I hypothesized that loss of one allele of
mTOR would reverse the differentiation arrest at the pro-myelin stage observed in Tsc1 cKO mice.
To assess this, I collected Dhh-Cre Tsc1 f/f and Dhh-Cre Tsc f/f mTOR f/+ mice and stained the
nerves for pro-myelin marker Oct6 and myelinating Schwann cell marker Krox20 (Figure 3A). I
found that the fraction of Oct6+ pro-myelinating cells decreased in P30 sciatic nerves (Figure 3B),
and the fraction of Krox20+ mature myelinating cells increased (Figure 3C) in Tsc1 cKO mice with
loss of one allele of mTOR when compared to Tsc1 cKO mice. However, the numbers of Oct6 and
Krox20+ cells were not fully restored to that of control mice.

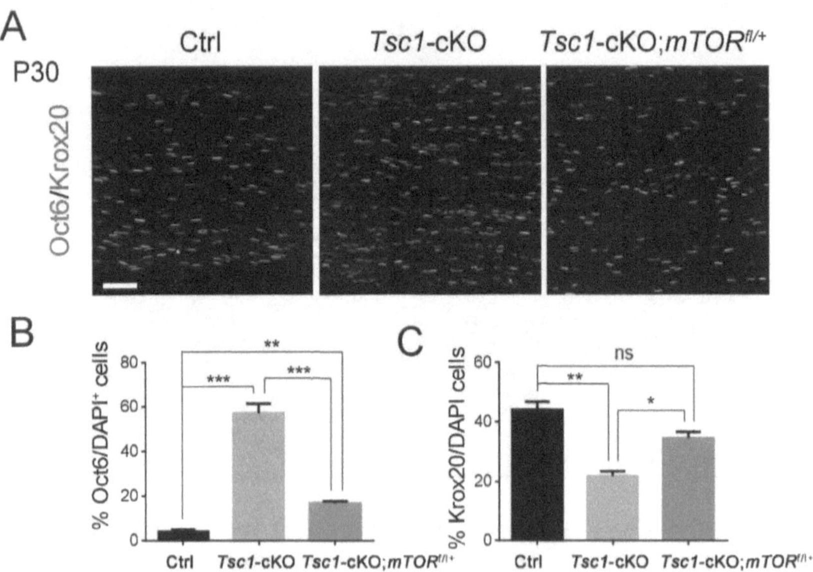

**Figure 3. Heterozygosity of mTOR Rescues Differentiation Defects that Result from Tsc1 Knockout in The
Schwann Cell Lineage.** A) Immunostained Oct6 and Krox20 in P30 sciatic nerves from control, Tsc1-cKO, and Tsc1-
cKO;mTORfl/+ sciatic nerves. Scale bar, 50 μm, B) Quantification of percentages of Oct6+ cells in control, Tsc1-cKO,
and Tsc1-cKO;mTORfl/+ sciatic nerves. n = 3 animals/genotype. One way ANOVA with Tukey's t-test used to
determine significance, C) Quantification of percentages of Krox20+ cells in control, Tsc1-cKO, and Tsc1-
cKO;mTORfl/+ sciatic nerves. n = 3 animals/genotype. One way ANOVA with Tukey's t-test used to determine
significance.

PLK1 Expression Decreases in Normal Schwann Cell Development and Is Elevated by mTOR Knockout in Mature Schwann Cells

Transcriptomic profiling of control and Tsc1 cKO nerves identified that genes in Polo-like kinase pathway were increased in Tsc1 mutant nerves. The preliminary data had shown increased expression of Plk1, Plk2, Cdc25b, and Cdc25c at the RNA level, and increased Plk1 at the protein level. From examination of this data, I noticed that Plk2 rather than Plk1 was the most increased gene in the Polo-like kinase pathway in Tsc1 cKO mice. To test to see if Plk2 was overexpressed at the protein level, I performed a Western blot of Tsc1 control and cKO samples using a commercial antibody that was available at the time. However, the bands were undetectable via Western blot, suggesting low protein expression of Plk2 or a bad antibody. Given the weakness of the Plk2 data, I did not pursue this further.

While the data that had been collected had demonstrated that Plk1 was increased at the RNA or protein level after Tsc1 knockout, a criticism that the reviewers raised was that these changes were due to artificial perturbations of Tsc1 rather than related to the natural process of Schwann cell differentiation. To explore how Plk1 levels change as Schwann cells decrease their mTOR activity and differentiate during the postnatal period, I performed qPCR on sciatic nerves for Plk1 at P0, P14 and P28. Consistent with our hypothesis that changes in Plk1 signaling help mediate the transition from proliferation to differentiation, I observed that Plk1 RNA level progressively decreased from P0 to P14 to P28, as developing Schwann cells differentiate from proliferating immature Schwann cells to mature myelinating Schwann cells (Figure 4A).

I next asked whether activation of mTOR in mature Schwann cells also led to the transcriptional upregulation of Plk1. I treated PLP-CreERT mice with tamoxifen and harvested sciatic nerves at P60 (Figure 4B). I found that PLK1 protein expression was upregulated in PLP-CreERT Tsc1 f/f mice, suggesting a conserved link between high mTOR activity and Plk signaling in Schwann cells (Figure 4C).

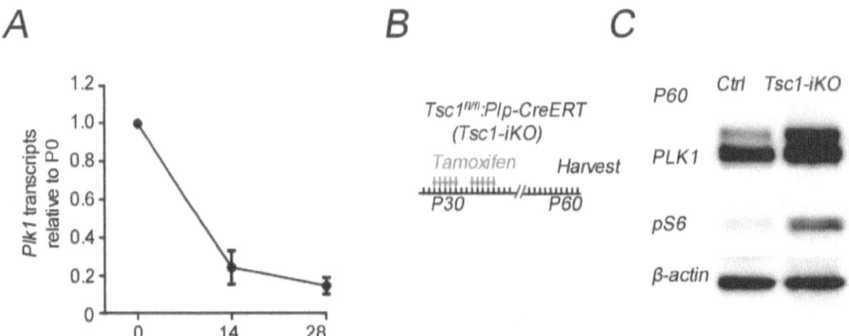

Figure 4. Dynamics of PLK1 During Normal Development and After Tsc1 Loss in Mature Schwann Cells. A) qRT-PCR of mRNA expression of PLK1 in WT sciatic nerves at P0, P14 and P28. n = 3 animals/group, B) Schema for knockout of Tsc1 in mature Schwann cells, C) Western blot of PLK1 and pS6 expression at P60 Tsc1-iKO mutants and control sciatic nerves after 10 days Tamoxifen injection at P30. β-actin was loaded as a control.

PLK1 Inhibition Induces G2 Accumulation and Impairs Developing Schwann Cell Proliferation

To target PLK1 signaling in vivo, we chose to use to use BI-2536, an inhibitor of Plk1, Plk2, Plk3 and Brd4 that has the highest activity against Plk1. To test the ability of BI-2536 to perturb cell cycle progression of Schwann cells, I performed two experiments. First, I seeded rat Schwann cells and treated them with DMSO or 100 nM BI-2536 for 24 hours. Rat Schwann cells treated with BI-2536 demonstrated disruption of the mitotic spindle and defects in chromosome segregation (Figure 5A). In addition to its role of regulating mitotic spindle formation, Plk1 promotes progression through the G2/M checkpoint by promoting activation of the cyclin B-Cdk1 complex[181]. To demonstrate whether prolonged treatment with BI-2536 could lead to accumulation of cells at G2, I performed cell cycle analysis using flow cytometry on rat Schwann cells 24 hrs after treatment with either DMSO or 100 nM BI-2536. As expected, I found an increase in the percent of Schwann cells at G2 (Figure 5B). I next asked whether BI-2536 would impair Schwann cell proliferation *in vivo*. Using previously collected samples that had either been treated with 5 mg/kg BI-2536 or vehicle for 7 days, I performed BrdU staining and analyzed the resulting data. Consistent with my findings with purified rat Schwann cells, I found that proliferation was impaired in Tsc1 cKO mice treated with BI-2536 when compared to vehicle treated mice (Figure 5C,D). My co-author concurrently found that BI-2536 treatment increased the number of Krox20+ mature myelinating Schwann cells.

Inhibition of PLK1 Signaling Rescues Hypomyelination induced by mTOR Hyperactivity

Finally, to determine whether Plk1 inhibition could reverse the hypomyelinating effects of mTOR hyperactivation in Tsc1 cKO mice, I performed a long-term treatment of Tsc1 cKO mice from P10-P21 with vehicle or BI-2536. I performed a Western blot to assess the expression of myelin basic protein and found that it was elevated when compared to Tsc1 cKO animals treated with vehicle (Figure 5E). I also collected sciatic nerves for ultrastructural analysis. Electron micrographs demonstrated visibly thicker myelin on P21 axons (Figure 5F). Quantification of these electron micrographs revealed an increase in the number of myelinated axons and a decrease in the g-ratio following BI-2536 treatment (Figure 5F,G).

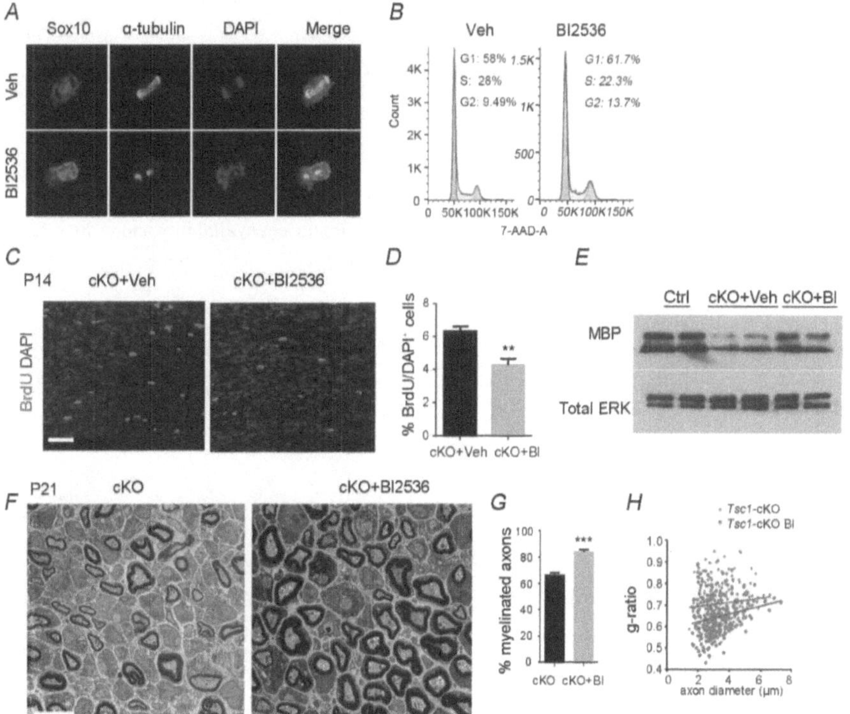

Figure 5. BI-2536 Rescues Myelination in Tsc1-cKO mice. A) Representative images of primary rat SCs treated with BI-2536 or vehicle and stained for Sox10, alpha-tubulin, and DAPI. Scale bar, 10 μm, B) Representative images showing cell cycle distribution of DMSO or 100 nM BI-2536 treated Schwann cells (N=2), C) Representative images showing BrdU staining of P14 Tsc1-cKO mutants treated with vehicle or with BI-2536 (N=3 animals per group). Scale

bar, 50 μm, D) Quantification of percentages of BrdU+ cells in Tsc1-cKO mutants treated with vehicle or with BI-2536. n = 3 animals/group. Student's t-test, ** p<0.01, E) Western blot showing MBP expression in representative P21 control treated with vehicle and Tsc1-cKO mice treated with vehicle or with BI-2536. Total Erk amount as a loading control, F) Representative electron micrographs showing ultrastructure of sciatic nerves in Tsc1-cKO mutants treated with vehicle or with BI-2536 at P21 (N=3 animals per group), G) Quantification of percentages of myelinated axons in Tsc1-cKO mutants treated with vehicle or with BI-2536 at P21 (n > 200 myelinated axons from 3 animals per group). Student's t-test, *** p<0.001, H) Quantification of g-ratio in Tsc1-cKO mutants treated with vehicle or with BI-2536 at P21 (> 200 axon counts from 3 mice per group).

Discussion

Myelination of the peripheral nervous system is critical for rapid conduction of action potential down nerve fibers and sensorimotor integration. Disruptions in myelin development or destruction of peripheral myelin leads to the formation of peripheral neuropathies such as Charcot-Marie-Tooth, Guillain Barre syndrome, and diabetic neuropathy. The myelinating cells of the peripheral nervous system, Schwann cells, develop from trunk neural crest, and differentiate based on microenvironmental cues from the extracellular matrix and their associated axon[16,172,174,175]. A developing area of interest is how Schwann cells integrate these cues to make cell fate decisions. The mTORC1 signaling pathway has been shown to integrate nutrient availability and growth factor signaling to promote cell growth[28]. Previous work done in Schwann cells had shown that loss of mTORC1 signaling in developing Schwann cells promotes hypomyelination[37,39]. The role of mTOR hyperactivity on Schwann cell development, in contrast, was not well understood. Hyperactivation of Akt, an upstream regulator of mTOR, can stimulate hypermyelination[43,45,191], but this finding is confounded by the fact that Akt can stimulate lipid synthesis through SREBPs independent of its effect on mTOR.

The work performed in this paper found that hyperactivation of mTOR following loss of Tsc1 led to hypomyelination in the peripheral nervous system. Similar to findings of other papers published around the same time, this hypomyelination was accompanied by an increase in Schwann cell proliferation and a block in Schwann cell differentiation at the pro-myelinating stage[184,189]. In contrast to other work which identified direct suppression of Krox20 by mTOR signaling[189], transcriptomic profiling identified the Polo-like kinase signaling pathway as a regulator of Schwann cell myelination. I built upon this work by demonstrating that Plk1 levels drop as Schwann cells differentiate to form mature myelinating Schwann cells. I then validated the role of the Plk1 pathway by demonstrating it impairs proliferation *in vitro* and *in vivo* and can rescue hypomyelination induced by mTOR hyperactivity. While other groups have noted an increase in cells expressing M-phase markers following Tsc1/2 knockout in Schwann cells[184], our work is the first to show a mechanism. Since our paper has been published, a study using Tsc2 null cells demonstrated more rapid G2/M progression following checkpoint arrest from DNA damage, mediated by increased PLK1 levels[192]. Higher levels of mTORC1 activity may plausibly be a mechanism that developing Schwann cells use to promote faster cell cycle transit times during early Schwann cell development. While decreasing cell cycle times has been shown to be a mechanism to regulate differentiation in the central nervous system[193], the same has yet to be demonstrated for myelinating glia of the peripheral nervous system. To address this question, we would perform dual EdU and BrdU labeling of sciatic nerves separated by 1hr, at four different time points of Schwann cell development (postnatal day 0, postnatal day 7, postnatal day 14 and postnatal day 28). At each time point, we would calculate the cell cycle length of developing Sox10+ Schwann cells and sees how it evolves over Schwann cell development both in normal mice and Tsc1 cKO mice.

The preliminary data showed that loss of one allele of mTOR could rescue the hypomyelinating effects of mTOR activity induced by Tsc1 loss. I built upon this finding by demonstrating that loss of one allele of mTOR partially rescued the differentiation arrest at the pro-myelinating stage induced by Tsc1 loss. Notably, loss of one allele of mTOR did not completely rescue the differentiation arrest at the pro-myelinating stage. This may be indicative of a more graded dose-response to the levels of mTOR signaling for Schwann cell differentiation. Alternatively, this difference may hint at mTOR independent effects of Tsc1 loss on Schwann cell differentiation. To examine the difference between these two hypotheses, we will obtain mTOR hypomorphic mice which show a 66% loss in mTOR expression[194,195] and breed with Tsc1 cKO mice. We would examine changes in sciatic nerve myelination, Schwann cell proliferation and Schwann cell differentiation in these mice to observe if they are improved with further reduction in mTOR signaling. To explore Tsc1-independent regulation of Schwann cells, we would perform mass spectrometry of control, Tsc1 cKO and Tsc1 cKO mTOR f/+ sciatic nerves. We would specifically be looking for a protein or phosphorylation event that is increased in Tsc1 cKO compared to control, but not affected by mTOR allelic reduction.

Preliminary data had shown that four-month-old Tsc1 cKO mice had formation of redundant myelin, hinting at a role for mTOR hyperactivation in postnatal myelin growth. I tested this idea directly by knocking out Tsc1 in mature Schwann cells using an inducible PLP-CreERT driver (Tsc1 iKO). I demonstrated that loss of Tsc1 in mature Schwann cells can lead to the formation of redundant myelin, independent of any effects it has on developing Schwann cells. This finding of myelin growth following Tsc1 knockout in mature myelin cells has been independently confirmed by two other groups[184,189]. Myelin growth after early development is driven by a combination of PI3K-Akt-mTOR signaling, which drives radial growth, and Yap/Taz activation from nerve stretching, which drives longitudinal growth[196]. Elevated levels of mTOR signaling in mature Schwann cells may disrupt this balance, leading to the formation of redundant myelin. My work found that levels of Plk1 were elevated in Tsc1 iKO mice when compared to matched controls. Surprisingly, Tsc1 iKO mice showed no changes in the number of Ki67+ cells, suggesting a cell cycle independent role of Plk1 signaling in Schwann cells. Tsc1 iKO sciatic nerves also displayed an increase in Oct6+ pro-myelinating cells. Oct6 is normally not expressed in mature Schwann cells, but has been shown to be induced by nerve injury[197]. During nerve injury, Schwann cells partially dedifferentiate in order to to remyelinate regenerating peripheral nerves[12]. Reactivation of mTORC1 signaling has been shown to be necessary for remyelination of sciatic nerves after crush injury[40]. Loss of Tsc1 in mature Schwann cells may therefore inappropriately activate genes associated with remyelination after injury in the resting state. While this process is deleterious in resting nerves, activation of mTORC1 signaling may represent a strategy to improve remyelination following nerve injury.

<u>Materials and Methods</u>

Animals

The floxed Tsc1 (Tsc1fl/fl) mice (Tsc1<tm1Djk>/J – 005680; Jackson Laboratory) were crossed with Dhh-Cre mice to produce control (Tsc1fl/+;Cre+/−) and Tsc1-cKO offspring. Tsc1fl/fl mice were crossed with Plp-creERT (Jackson Laboratory, stock #005975) mice to produce control (Tsc1fl/+;Plp-CreERT+/−) and Tsc1-iKO offspring. Animals of both sexes were used in the study and heterozygous littermates were used as controls unless otherwise indicated. The mouse strains used in this study were generated and maintained on a mixed C57Bl/6;129Sv background and housed in a vivarium with a 12-hour light/dark cycle. All animal use protocols and studies were approved by the Institutional Animal Care and Use Committee of the Cincinnati Children's Hospital Medical Center.

Histology and electron microscopy

The sciatic nerves of mice at defined ages were dissected and fixed overnight in 4% paraformaldehyde and processed for cryo-sectioning or fiber analyses. Tissue processing was performed essentially as described previously (Wu et al. 2016). Briefly, mice were deeply anesthetized with ketamine/xylazine, perfused with 0.1 M cacodylate, followed by 2.5% paraformaldehyde/2.5% glutaraldehyde in 0.1 M cacodylate (pH 7.2). Sciatic nerves were dissected, post-fixed in 1% OsO4, dehydrated through a graded ethanol series, infiltrated in propylene oxide, and embedded in resin. Semi-thin sections were stained with toluidine blue, and ultrathin sections were stained with lead citrate.

Immunohistochemistry and immunoblotting

Cryosections (8-μm thick) were permeabilized and blocked in blocking buffer (0.3% Triton X-100 and 5% normal donkey serum in PBS) for 1 h at room temperature and overlaid with primary antibodies overnight at 4 °C. Antibodies used in the study were goat anti-Sox10 (Santa Cruz Biotechnology, sc-17342), rabbit anti-Sox10 (Abcam, ab-25978-1000), goat anti-Oct6 (Santa Cruz Biotechnology, sc-11661), rabbit anti-Egr2 (Santa Cruz Biotechnology, sc-20690), goat anti-MBP (Santa Cruz Biotechnology, sc-13914), TSC1 (Pierce PA5-18506), Ki67 (Thermo Scientific, RM9106-SO), rat anti-BrdU (Abcam, ab6326), mouse anti-PLK1 (Santa Cruz Biotechnology, sc-17783), goat anti-Sox2 (Santa Cruz Biotechnology, sc-17320), P-S6 (Cell Signaling Technology, 2211), S6 (Cell Signaling Technology, 2217), p-4EBP1 (Cell Signaling Technology, 9451), α-Tubulin (Cell Signaling Technology, 2144) and mouse anti-GAPDH (Millipore, Mab374). After washing in PBS, cells or sections were incubated with secondary antibodies conjugated to Cy2, Cy3, or Cy5 (Jackson ImmunoResearch Laboratories) for 2 h at room temperature, stained in DAPI for 10 min, washed in PBS, and mounted with Fluoromount-G (SouthernBiotech). Cell images were quantified in a double-blinded manner.

For BrdU incorporation, 100 mg/kg BrdU were injected by IP. Two hours later, mice were harvested and processed for immunohistochemistry analysis. For immunoblotting, sciatic nerves were incubated in 1× Passive Lysis Buffer (Promega) supplemented with a protease inhibitor cocktail (1:200, Sigma). After western blotting, proteins were detected with appropriate secondary antibodies by using chemiluminescence with the ECL kit (Pierce) according to the instructions of the manufacturer.

PLK inhibitor treatment

PLK inhibitor BI-2536 (Biochempartner, 755038-02-9) was dissolved in ethanol and stored at a stock concentration of 10 mg/ml in aliquots at −20 °C. The working solution was prepared freshly before use at a final concentration of 1 mg/ml in 4% ethanol, 5% Tween 80, and 5% PEG400. Mice were administered daily intraperitoneal injections of either BI-2536 (5 mg/kg), as previously reported for in vivo treatment (Oueslati et al. 2013), or vehicle once per day from P7 to P14 or from P10 to P21. Mice were then harvested and analyzed by immunohistochemistry and ultrastructure analysis. For primary rat Schwann cell treatment, 100 nM BI-2536 working solution was mixed in the differentiation medium.

RNA extraction and qRT-PCR

Analyses were conducted with RNA extracts from a pool of sciatic nerve tissues of P0, P7 or P14 mutant mice and their littermate controls. Total RNA was extracted per the Trizol (Life Technologies) protocol. cDNA was generated with iScript™ cDNA Synthesis Kit (Bio-Rad). qRT-PCR was performed using the ABI Prism 7700 Sequence Detector System (Perkin-Elmer Applied Biosystems). qRT-PCR primers for mouse gene sequences were PLK1-f, CAGCAAGTGGGTGGACTATT; PLK1-r, AGAGAATCAGGCGTGTTGAG; PLK2-f, GAGGACAGGATCTCTACAACTTTC; PLK2-r, AGAGCATGTTCAGGGCATATT; CDC25A-f, GACCAGTATTGCTGCTACTCAA; CDC25B-r, GGTCTGGGAAGGTTAGCTTATG; CDC25B1-f, CACCTCTCGGTCTTTGAGTTT; CDC25B-r, TGTGCATGGTCTGTGTAAGAG; CDC25C-f, CATTCAGATGGAGGAGGAAGAG; CDC25C-r, CACTGTGTCTGGGCTGATATAC; Cdk1-f, CTGTTTGGAGGATCTCGGTAAG; Cdk1-r, TTCCCTGACTCCAGCAAATG; GAPDH-f, ACCCAGAAGACTGTGGATGG; and GAPDH-r, CACATTGGGGGGTAGGAACAC.

Cell Cycle Analysis

One million purified rat Schwann cells per group were seeded onto 10 cm poly-L-lysine (PLL)-coated dishes and allowed to attach overnight. Cells were then treated with DMSO or 100 nM BI-2536 for 24 hrs. Cells were collected, fixed using cold 70% ethanol and stained with 7-AAD (7-amino actinomycin D). Stained cells were run on a BD™ LSR II flow cytometer and gated on forward and side scatter. Cell cycle distributions were computed using FlowJo platform (https://www.flowjo.com/)

CHAPTER 2: Glioblastoma Genetic Drivers Dictate the Function of Tumor-Associated Macrophages/Microglia and Responses to CSF1R Inhibition

Introduction

Glioblastoma (GBM) is a highly malignant brain tumor with poor prognosis. Despite aggressive multimodal therapies including radiation and temozolomide [198], median survival of GBM patients is a dismal 14 months. Based on transcriptional profiling, human GBMs have been classified into at least three subtypes including a proneural subtype associated with oligodendrocyte lineage signatures that often exhibits *PDGFRA* amplification, *TP53*, loss and increased phosphoinositide 3-kinase (PI3K) signaling; a classical subtype enriched in astrocytic signatures characterized by EGFR gain of function; and a mesenchymal subtype associated with immune infiltration and the loss of *NF1* and *TP53* [57,58]. However, clinical trials targeting subtype-specific signaling pathways in GBMs have not been successful [50-52].

Individual tumors display significant molecular and cellular heterogeneity with diverse co-existing cell types including tumor cells, immune cells, and vascular cells [73,133]. Single-cell profiling has demonstrated that individual tumor cells exist on a continuum between four cell lineages: astrocyte-like, oligodendrocyte progenitor cell-like, neural progenitor cell-like, and mesenchymal-like subtypes [77]. Brain tumor cells have been shown to recruit immune cells and vascular networks to support their growth [133]. How the tumor microenvironment (TME) influences the growth of genetically and phenotypically distinct glioma subtypes remains poorly understood. Therapies targeting the TME represent a potential approach to anti-glioma therapy due to the mutational stabilities of these cells[133]. Tumor-associated macrophage/microglia (TAMs) are the most prevalent microenvironmental cells in GBM [133], especially in the mesenchymal subtype of GBM[57]. TAMs in the central nervous system (CNS) include microglia, CNS border-associated macrophages, and monocyte-derived macrophages[110]. Resident microglia in the brain play important roles in brain development, homeostasis, maintenance of neuronal networks, and local inflammation [112,113]. Border-associated macrophages present at CNS interfaces, such as the perivascular and perimeningeal spaces [113], participate in antigen presentation to infiltrating T cells [114]. Monocyte-derived macrophages localize to the tumor core of mouse and human gliomas, support angiogenesis and tumor growth, and are associated with poor prognosis in GBM [115-117]. Within human gliomas, two additional TAM subtypes have been identified: a CD73[high] immunosuppressive subset [127] and a MARCO[high] subset of TAMs that promote mesenchymal transition and aggressive tumor behaviors [199].

Targeting of TAMs has been explored through evaluation of small-molecular inhibitors of macrophage colony-stimulating factor-1 receptor (CSF1R) signaling, which is involved in recruitment, proliferation, and polarization of TAMs [133]. CSF1R inhibition in mouse models of glioma has revealed that TAMs function in tumor growth [118,119,121,122,134,135], immunosuppression [200], regulation of vascular networks [103] and cell invasion [201], depending on the tumor model. Treatment of a mouse PDGFB-driven glioma model with CSF1R inhibitor BLZ945 promoted durable regression and dramatically improved long-term survival of tumor-bearing mice [134]. Inhibition of CSF1R signaling has been linked to increased responses to tyrosine kinase inhibitors in a PDGFB-driven model of glioma [136]. However, studies in a syngeneic GL261 glioma model showed mixed results that CSF signaling inhibition either accelerated or decreased tumor growth[103,202]. Clinical trials of CSF1R inhibitor PLX3397 in recurrent human GBMs showed no population-wide response, while two patients in the trial showed an extended survival response, suggesting that glioma subtypes vary in their responses to CSF1R-dependent TAM targeting [137].

To determine the responsiveness of glioma subtypes to TAM targeting, we examined different animal models of glioma subtypes for sensitivity to CSF1R inhibition. Similar to the previous report [136], we found that the CSF1R inhibitor hindered tumor growth in a proneural-like PDGFB-DNp53-induced glioma model. In contrast, CSF1R inhibition in oncogenic RAS-driven mesenchymal-like gliomas led to acceleration of tumor growth at the early phase of tumor progression. Single-cell transcriptomic profiling revealed that the programs associated with inflammation and immunosuppression are upregulated in TAMs within the microenvironment of RAS-driven tumors compared with those in the PDGFB-driven TME. In RAS-driven gliomas, interactions between TAMs and the vasculature are prevalent. Further, co-targeting of TAM and angiogenesis reduced proliferation and tumor mass in RAS-driven gliomas. Together, our studies suggest that the microenvironment landscapes elicited by molecularly distinct GBM subtypes dramatically impact the efficacy of anti-microenvironment therapies.

Methods

Cloning of HRasV12-dnp53 Viral Vector

The pQCXIX-PDGFB-IRES-dnp53 vector was digested with NotI and EcoR1 (New England Bioloabs R1389S, R3101S, Ipswich, MA) to remove the PDGFB insert. Human HRas G12V was amplified via PCR using Phusion polymerase (NEB M0530S) from pWZL-hygro-HRasV12 (Addgene #18749) and ligated into the pQCXIX-IRES-dnp53 fragment using T4 Ligase (NEB M0202S). Successful integration was confirmed by Sanger sequencing performed by the DNA Sequencing Core.

Plasmid purification

For small scale purifications, DNA was purified using the Qiagen Miniprep kit (Qiagen 27104, Hilden, Germany). For packaging of retroviral stocks, plasmid was purified using a cesium chloride gradient extraction. In brief, plasmid-containing Stbl3 bacteria were expanded overnight in 200 mL culture. The next day, bacteria were pelleted and nucleic acids were extracted from bacteria through alkaline lysis and were precipitated using isopropanol. Nucleic acids were resuspended in Tris-EDTA buffer with cesium chloride and ethidium bromide added. The solution was then placed in a Beckman Coulter Optiseal Tube (Beckman Coulter 361623, Brea,CA) and centrifuged in a 70.1Ti rotor at 40,000 rpm for 18.5 hours. The next day, the DNA containing band was extracted using a 17 g needle and transferred to a fresh tube. The band was washed 5 times with water saturated butanol 5 times to remove ethidium bromide from the solution and the DNA was pelleted with 100% ethanol. The DNA pellet was allowed to resuspend in 10 mM Tris-EDTA overnight. The next day, DNA was pelleted again using ethanol/sodium acetate. After a wash with 70% ethanol, the DNA pellet was resuspended in TE. The concentration was quantified with a Nanodrop spectrophotometer

Virus Production

From a confluent dish, 293T cells were split 1:6 into new dishes. The next day, the cells were transfected using the calcium-phosphate method with 7.5 ug of transfer plasmid (pQCXIX-PDGFB-IRES-dnp53 or pQCXIX-HRasV12-IRES-dnp53) and 7.5 ug of pCL-Eco (Addgene #12371). The next day, media was changed to fresh media. Forty-eight hours after the media change, supernatants were collected, spun down at 1200 rpm to remove cellular debris and was filtered with a Corning 0.22 uM Bottle Top Vacuum Filter (Corning 430773, Corning, NY). The filtrate was then loaded into ultracentrifuge tubes (Beckman Coulter 344058, Brea, CA) and was spun at 25000 rpm in a SW28 rotor for 2 hr. Supernatant was discarded and pellet was resuspended overnight in 50 uL of HBSS at 4C. The next day, the HBSS-virus solution was aliquoted into 10 uL aliquots and kept at the -80 C until the day of injection.

Stereotactic Injections

Mice were anesthetized using 2% isoflurane and loaded into onto a stereotaxic device (RWD Life Sciences, San Diego, CA). The skin on the skull was disinfected using isopropanol and chlorohexidine gluconate and was incised to expose the skull. From the bregma, the syringe was moved to the appropriate AP and lateral coordinates, and a small hole was drilled using a microdrill (RWD Life Sciences 78001, San Diego, CA). The Hamilton 1701 syringe (Hamilton Company, Reno, NV) was lowered to the level of the dorsal-ventral coordinate and a QSI automated injector (Stoelting

53312, Wood Dale, IL) was used to inject 2 uL over 5-6 min at a rate of 0.3 uL/min. After the injection was completed, the needle was allowed to sit for 2 min in the brain before it was carefully removed. The mouse's head wounds were sealed up with Gluture and they were allowed to recover in an incubator. To deal with post-operative pain, mice received 40 uL of 0.05 mg/mL Buprenex.

Mice were injected with either cells or virus. For virus, viral aliquots were taken from -80 freezer and thawed on ice. To each aliquot, polybrene (Millipore Sigma, St Louis, MO) was added to a final concentration of 1 mg/mL. Virus was injected in the subventricular zone (-1.5, 2.0, -2.3) or subgranular zone (-2,1.5,-2.3). For cell injections, cells were dissociated using Accutase (Stem Cell Technology 07922, Vancouver, CA) and resuspended in their base media to a final concentration of 50,000 cells/uL. Cells were either injected into the cortex (1, 0.5, -2) or the striatum (for PDGFB cell lines, coordinates (0.2, 2.2,-3).

Cell Culture

293T cells were cultured in DMEM with 10% FBS with 2 mM L-glutamine and penicillin/streptomycin at 37 C. When passaging these cells, media was removed and cells were washed with DPBS. 1 mL of 0.05% Trypsin-EDTA (ThermoFisher 25300120, Waltham, MA) was added and was allowed to incubate for 5 min. Trypsin was neutralized with addition of DMEM and were move to a new dish.

Human PDX lines GSC262 and GSC20, and Ras cell lines were grown in serum free DMEM/F12 media supplemented with 60 uM putrescine, 2 ug/mL heparin, 20 nM progesterone, 0.1% BSA, 1X ITS supplement, 5 mM HEPES, 0.27% glucose and 1% Penicillin/Streptomycin. To base media, EGF and FGF (Shenandoah Biotechnology, Warwick, PA) were added to final concentrations of 20 ng/mL.

Mouse PDGFB-dnp53 and PDGFRA-D842V-dnp53 cell lines were grown on low attachment dishes. Low attachment dishes were made by adding 3 mL of Anti-Adherence Rinsing Solution (Stemcell Technologies, Vancouver, CA) to a standard 10 cm dish and incubating at room temp for 5 min. The solution was then removed and the coated dish was washed twice with DPBS before adding media. These cells were cultured in TSM media. Base TSM media was comprised of a Neurobasal/DMEM-F12 mixture (1:1) with 2 mM Glutamax, 1X Non-Essential Amino Acids, 1 mM sodium pyruvate and 1 mM HEPES added. Complete media was made from base media by adding B27 and N2 supplements, heparin, rhEGF, rhFGF, rhPDGFBB, and rhPDGFAA (Shenandoah Biotechnology, Warwick, PA)to a final concentration of 20 ng/mL.

Dissociation of cells in serum free media was performed by collecting the media with suspended cells, spinning down for 5 min at 1200 rpm to pellet the cells, removing the supernatant and resuspending in 1mL Accutase. Cells were kept in the incubator for 5 min and then fresh media was added to neutralize the enzymes. Cells were moved to a new dish.

Stably infected luciferase lines were derived by infecting cells with 2uL CMV-firefly-luciferase lentivirus puromycin (Cellomics Technology, Halethorpe, MD) per 200,000 cells. Stably transduced clones were selected using 2 ug/mL puromycin (Invivogen, San Diego, CA).

<u>Derivation of primary tumor cell lines</u>

Mice were anesthetized using isoflurane and were perfused in the left ventricle with ice cold PBS. Tumors were dissected out from the brain and were minced finely with safety blades. The tumor suspension was then transferred into 2 mL of Accutase and was allowed to incubate for 20 min at 37 C with mild agitation. Media was added to neutralize the Accutase and the suspension was filtered using a 40 uM filter. Cells were spun down at 1200 rpm for 5 min and were resuspended in Hybridimax Red Blood Cell Lysing Buffer (Millipore Sigma R7757, St. Louis, MO). After 1 min, 10 mL of DPBS was added to neutralize lysis buffer and cells were pelleted. Cells were resuspended in complete media (normal or TSM serum free media) and were placed in the 37 C incubator.

<u>Cell Viability Assays</u>

Tumor cells were seeded to 20,000 per well in a 96 well plate. For each dose level of drug, there were three wells. Cells were incubated in drug or vehicle control for 3 days. On the third day, WST-1 (Takara MK400, Kusatsu, Japan) was added and incubated for 2 hours per the manufacturer's protocol. The absorbance at 405 and 670 nM was then read per well using a SpectraMax M2 spectrophotometer (Molecular Devices, San Jose, CA)

<u>Animal Experiments</u>

For our experiments we used Boy/J mice we bred in house or purchased from the CCHMC Comprehensive Mouse and Cancer Core. Both genders of mice were used. PLX3397 hydrochloride (Medkoo Biosciences, Cary, NC), BKM120 (Medkoo Biosciences, Cary, NC) and Cediranib free base (LC Laboratories, Woburn, MA) were used for drug treatments via oral gavage in vivo. PLX3397 was first dissolved in DMSO and was then added to a solution of 1% Tween-80 (Millipore Sigma, St Louis, MO) and 0.5% hydroxypropylmethylcellulose (Millipore Sigma, St Louis, MO) in water to make a working solution for mouse gavage. Light sonication was performed to disperse the

PLX3397 evenly. Cediranib was dissolved in 1% Tween-80 for single treatments and 1% Tween-80/0.5%HPMC for combination therapy. PLX3397 was administered at 100 mg/kg via oral gavage daily starting on either day 1 after tumor cell transplantation or when the tumor was established (7 or 14 days depending on the tumor model). Cediranib was dosed at 6 mg/kg via oral gavage starting at 7 days after tumor cell transplantation. BKM120 was first dissolved in N-methyl-2-pyrrolidone (Millipore Sigma, St. Louis, Missouri) at a stock concentration of 40 mg/mL and stored in the -80. On the day of gavage, BKM120 aliquots were diluted 1:10 in PEG300 (Millipore Sigma, St. Louis, MO) and were administered to the mice at a dose of 20 mg/kg daily. Mice receiving both PLX3397 and BKM120 received their dose of PLX3397 in the morning and their dose of BKM120 in the evening.

When mice reached a desired timepoint or became moribund, they were sacrificed by anesthesia with isoflurane and perfusion with ice cold PBS. Mice whose tissues were collected for RNA collection had tumors dissected out and frozen on dry ice. Mice destined for histology were perfused with 4% PFA in PBS, followed by a post-fix in 4% PFA overnight and a change to PBS.

Bioluminescent Imaging of Mice

For bioluminescent imaging of mouse tumors, mice received an intraperitoneal injection of 100 uL of 30 mg/mL potassium D-luciferin solution (Gold Bio, St Louis, MO) dissolved in DPBS. The mice were anesthetized using isoflurane and loading onto an IVIS Spectrum CT imaging platform (Perkin Elmer, Waltham, MA). At 10 minutes, mice were imaged with IVIS with an exposure time of 10 seconds- 3 minutes. The images were than scaled using a lower bound of 150 counts and an upper bound of 65000 counts. Total flux was calculated by drawing a rectangular ROI around the mouse's head and measuring the radiance in that ROI from the Living Image software.

Vibratome sectioning and staining

Mouse brains were cut sagittally and were embedded in 4% agarose. Brains were then sectioned using a Leica VT1000S vibratome with 50 uM sections. For staining, sections were transferred to a 24 well plate and blocked with 5% NDS in PBS 0.1% Triton-X for 1 hr. The blocking solution was then removed and a primary antibody solution with 5% NDS in PBS was added. The primary antibodies were incubated overnight at 4 C. The next day, the sections were washed 4X10 min with PBS. Secondary antibody solution in 5% NDS/PBS was then added and incubated for one hour at room temperature. DAPI solution of 1 ug/mL in PBS was then added to replace the secondary solution and was incubated for 10 min. Sections were then washed 4X10min in PBS and then carefully loaded onto a glass slide. Fluoromount-G Mounting media was then added to the sections (Southern Biotech 0100-01, Birmingham, AL) and the slides were kept under dark before imaging.

<u>Paraffin Sectioning and Staining</u>

Sagittally cut brains were submitted to the CCHMC Pathology Core for embedding in paraffin. Once embedded, samples were cut into 5 uM sections with a Leica RM2235 microtome. Prior to staining, slides were placed in a 65 C incubator overnight to melt off excess wax. Samples were then deparaffinized through sequential changes of xylenes, 100%, 95% and 75% ethanol and moved to PBS. Samples were then put in antigen retrieval solution (10 mM sodium citrate/citric acid buffer, pH6.0) in a steamer for 35 min. Solutions were allowed to cool down to room temperature and then were washed with PBS for 5 min. They then were blocked for 2 hours in 5% NDS, 0.1% Triton-X. Primary antibodies added in 5% NDS/PBS were incubated overnight. The next day, slides were washed 5X5min in PBS and fluorescent secondary antibodies were added at a dilution of 1:500. Secondary antibodies were incubated for 1 hr, followed by counterstaining with 1 ug/mL DAPI. After another series of washes in PBS 5x5 min, the slides were mounted with Fluoromount-G and stored under dark until imaging.

For immunohistochemistry staining, paraffin was melted off, slides were deparaffinized and underwent antigen retrieval as previously described. Following antigen retrieval, slides were washed in PBS 2X5 min and then blocked with 0.3% hydrogen peroxide in 0.2% Triton-X-100/PBS for 10 min. Slides were washed in PBS 3X5min and then transferred to 5% normal goat serum (NGS) in 0.2% Triton-X-100/PBS and incubated for 1 hr. Serum blocking buffer was removed and primary antibody added in 5% NGS overnight. The next day, primary antibody was removed in slides were washed in PBS 3x5min. Biotinylated secondary antibody was added in 5% NGS/PBS for 1hr. Slides were washed with PBS and PBST and then ABC reagent (Vector Labs PK-600, Burlingame, CA) was added for 1 hr for signal amplification. Slides were then incubated in 0.05% diaminobenzene, 0.015% hydrogen peroxide and 0.01M PBS to develop the signal for 3 min. Slides were transferred to water to stop the reaction and were counterstained with hematoxylin. Slides were mounted using Cytoseal mounting media (Richard Allen Scientific, San Diego, CA).

<u>Antibodies</u>

We used the primary antibodies Rabbit anti-phospho-H3 S10 (Cell Signaling Technology, 9701S) 1:300, Rabbit anti-phosho-S6 S235/S236 1:300 (Cell Signaling Technology, 2217S), Rabbit anti-CD31 1:200 (Abcam, ab28364), Rabbit anti-Cleaved Caspase 3 Asp175 (Cell Signaling Technology, 9661S), Rabbit anti-Iba1 1:500 (Fujifilm Wako Chemicals USA, 4987481428584), Rabbit anti-M-CSF1R Y723 49C10 1:200 (Cell Signaling Technology, 3155S), Rabbit anti-pAkt S473 (Cell Signaling Technology 3787, 1:100) and Goat anti-Sox2 1:100 (Abcam ab239218).

For secondary antibodies, we used Donkey anti-Rabbit Alexa 488 (Jackson Immunoresearch, 711-545-152), Donkey anti-Goat Alexa 488 (Jackson Immunoresearch, 705-545-147), Donkey anti-Rabbit Cy3 (Jackson Immunoreseach, 711-165-152), Donkey anti-Rat Cy3 (Jackson Immunoresearch, 712-165-150), and Donkey anti-Rabbit Cy5 (Jackson Immunoresearch, 711-175-152). All antibodies were given at a dilution of 1:500. For IHC secondary, we used Goat anti-Rabbit Biotin 1:500 (Jackson Immunoresearch, 111-065-003)

<u>Vessel Quantification</u>

Vessel quantification following anti-angiogenic therapy was performed using Angiotool[203] with a vessel diameter set to 10, and other parameters sent do defaults. All other quantifications were done using the multipoint tool in ImageJ.

<u>RNA Extraction and Sequencing</u>

Samples were homogenized in Trizol and nucleic acid fraction was separated using phenol-chloroform extraction and Phase-Lock-Gel Heavy Tubes (VWR, 10847-802). RNA was precipitated by ethanol extraction and incubation on ice and pelleted through centrifugation. RNA was then resuspended in DEPC-Water and treated with DNAse I for 10 min. Sample was then purified using a RNeasy column (Qiagen 74104, Hilden, Germany). The RNA concentration was measured using a Nanodrop spectrophotometer. For RNA sequencing, 1 ug of purified RNA sample was sent to BGI Americas to undergo DNB-seq.

<u>Processing of Bulk Transcriptomic Data</u>

Alignment of RNA-seq reads to the mm10 reference genome was performed by the CCHMC Bioinformatics Core. RNA-seq reads in FASTQ format were first subjected to quality control to assess the need for trimming of adapter sequences or bad quality segments. The programs used in these steps were FastQC v0.11.7[204], Trim Galore! v0.4.2[205] and cutadapt v1.9.1[206]. The trimmed reads were aligned to the reference human genome version GRCm38/mm10 with the program STAR v2.6.1e[207]. Aligned reads were stripped of duplicate reads with the program sambamba v0.6.8[208]. Gene-level expression was assessed by counting features for each gene, as defined in the NCBI's RefSeq database[209]. Read counting was done with the program featureCounts v1.6.2 from the Rsubread package[210]. Raw counts were normalized as transcripts per million (TPM).

<u>Analysis of Bulk Transcriptomic Data</u>

Differential expression analysis was performed using the DESeq2 package from R 4.0.2 using raw gene counts as an input. Significant genes were defined as those with FDR<0.05 and a log2 fold change of at least 2. Heatmaps were drawn using the pheatmap package from R. For gene set enrichment analysis, normalized counts were generated using the counts function in DESeq2 and were used as an input for GSEA 4.0.3.

<u>Dissociation of Tissues for Single-Cell RNA-seq and Sequencing of Libraries</u>

Primary mouse tumors were dissected out of the mouse brain and minced finely with a pair of safety blades. The cell suspension was then added to a solution of 2 mL collagenase IV and incubated at 37 C for 20 min with mild agitation. Collagenase was neutralized following the addition of serum free media, and cell suspension was passed through a 40 uM filter. Cells were pelleted and resuspended in RBC Lysis Buffer (Sigma, 11814389001). After one minute of incubation, buffer was neutralized using PBS and cells were pelleted again using centrifugation. Cells were resuspended in 0.01% BSA and were submitted to the CCHMC Gene Expression Core for either Drop-Seq or 10x Chromium 3v3 Profiling. 10X Libraries were sequenced using the Novo-seq platform at CCHMC DNA Sequencing Core. Drop-Seq libraries were sequenced using the Center for Genomics and Bioinformatics at Indiana University and Cincinnati Children's Hospital Medical Center using a MiSeq platform. Gene barcode matrix files were generated from fastq files using the cellranger count method.

<u>Visualization and Analysis of Single-Cell RNA-seq Data</u>

Cells were analyzed using and visualized using the Seuratv3 package in R. We filtered low quality cells out, which we defined as those having less than 200 or more than 50,000 reads, or a mitochondrial percentage greater than 50%. We used the harmony package[211] to normalize batch effects between different datasets, treating each sample as its own batch. Based on the ElbowPlot function, we chose around 43 principal components for UMAP driven visualizations. Clustering was performed using the FindNeighbors and FindClusters functions in Seurat. Additionally, if the UMAP visualization showed a cluster of cells not captured by standard methods, the CellSelector command was used to define additional clusters. Markers for each cluster were defined from a combination of literature knowledge and the FindMarkers function in Seurat. Differential expression analysis was also performed using the FindMarkers function. Gene set enrichment analysis on single cell data was performed using the msigdbr, presto[212] and fgsea packages. Heatmap visualization was performed using the pheatmap package, and Rich plots were generated as previously described[213].

We used the CellChat package to perform receptor-ligand analysis on our tumors.[214] For identifying significantly expressed genes in each cluster, we used the truncatedMean option, choosing genes that were expressed in at least 10% of the cells, and told the program to consider population size in its determination of ligand importance.

We downloaded the following published datasets for single cell RNA-seq analysis from GEO: GSE138794[78], GSE131928[77], GSE103224[215], and GSE84465[216].

Defining Mouse Macrophage Signatures

Mouse macrophage signatures were defined by comparing Ras BAMs to PDGF Microglia and selecting the top 10 differentially expressed genes. Scoring of these signatures in Human gliomas was calculated by performing one to one mapping of mouse genes using the biomaRt package, and summing the expression of all the signature genes.

Glioma Subtype Calculation

Subtype calculation was performed using a previously described method[73]. Subtype signature genes for glioma were collected and initial scores were calculated by summing the expression of all subtype specific signature genes and subtracting the average gene expression for that cell.[57] To determine the significance of these scores, a distribution of scores was generated by randomly choosing gene sets of the same size as the number of genes in the subtype gene set and calculating a score as shown above. From this set of 4000 scores, a 5% percentile cutoff was calculated using the quantile function in R. Subtype scores that fell above this threshold were deemed significant. The highest significant subtype score was used to assign the identity of the cell. If not subtype scores were significant, the cell was defined as a mixed identity.

Statistics

All quantification plots were generated in Prism 9.0. Survival curves were evaluated for significance based on the log rank t-test. For quantifications of images, either ANOVA with multiple comparisons test was used to determine significance (for 3 or more groups), or an unpaired t-test with Welch's correction (2 groups).

Results

Targeting TAMs blocks tumor growth in murine PDGFB-driven gliomas but not human proneural GBMs

To determine the function of TAMs in glioma growth, we isolated primary tumor cells from an animal model of proneural GBM [217] that overexpressed human PDGFB and dominant-negative p53 (dnp53). We implanted these cells in immuno-competent adult mice to establish a syngeneic GBM model in which recipient mice develop full-blown tumors at around 14 days post implantation. Recipient tumors exhibit pathological and morphological characteristics of human GBM [217].

To inhibit macrophage activity, we treated mice daily with PLX3397, an FDA-approved oral brain-penetrant CSF1R inhibitor that causes macrophage depletion or reprogramming [134,135], starting the day after tumor implantation (Figure 1A). In contrast to vehicle-treated mice that died around 32 days, the treatment with PLX3397 blocked the growth of the transplanted tumor cells and significantly ($p = 0.0001$) extended animal lifespan (Figure 1B, C). There were no apparent tumor masses in the PLX3397-treated mice that had been transplanted with PDGFB-driven GBM cells (Figure 1D). Mice that died while being treated showed small residual tumors with high degrees of hemorrhage (Figure 1D).

Examination of tumor tissues after three days of PLX3397 treatment showed a dramatic decrease in the number of proliferative phospho-H3$^+$ cells and in CD31$^+$ blood vessel density (Figure 1E, F). Despite the reduction of CSF1R activity, the number of Iba1$^+$ TAMs was not altered by PLX3397 treatment (Figure 1G). This suggests that TAMs are reprogrammed rather than depleted by PLX3397 in the PDGFB-driven GBM model. This is consistent with a previous report in a similar PDGFB-induced glioma model (RCAS-hPDGFB/Nestin-Tv-a;Ink4a/Arf$^{-/-}$) [134,135].

We next assessed whether targeting TAMs with the CSF1R inhibitor extended survival in established tumors. We treated established PDGFB-driven GBM-bearing mice with PLX3397 at 14 days after tumor transplant (Figure 1H). We found that PLX3397 treatment extended mouse survival

(Figure 1I) and inhibited the progression of established tumors (Figure 1J). In addition, in contrast to the strong effect of PLX3397 on tumor growth *in vivo,* we found that treatment of tumor cells with PLX3397 *in vitro* had no effect on cell viability (Figure 1K). These observations suggest that CSF1R inhibition targets TAMs rather than tumor cells to block initiation and progression of PDGFB-driven GBM tumors.

To investigate if TAMs are important in the growth of human proneural GBMs, we used a patient-derived xenograft (PDX) GBM model transplanted with proneural glioma stem-like cells (GSC262) [218] – a cell line with *EGFR* amplification, *TP53* mutation, and *CDKN2A* deletion [219]. We treated these mice with PLX3397 after the tumor had been established in the brain. In this model, CSF1R inhibition did not have significant effects on tumor growth, histology, or animal survival (Supplementary Figure S1). This suggested that, unlike the proneural-like PDGFB-driven syngeneic mice, human proneural GBM PDX tumors derived from GSC262 are not responsive to CSF1R inhibition.

RAS-driven mesenchymal-like gliomas are resistant to CSF1R inhibition

GBM tumors of the mesenchymal subtype are driven by activation of RAS signaling due to *NF1* loss and are densely infiltrated by TAMs [57]. To determine whether CSF1R inhibition could suppress tumor growth of mesenchymal GBM, we developed an animal model of mesenchymal-like gliomas by expressing oncogenic human HRASV12 and dnp53, which mimic the loss of *NF1* and *TP53* (designated as RAS-driven GBM). Expression of HRASV12 and dnp53 has previously been shown to induce the mesenchymal-like GBM in mice [220]. Adult mice transduced with HRASV12-dnp53 retrovirus in the subventricular zone developed GBMs as early as 30 days post-transduction (Supplementary Figure S2A, B). Secondary tumors derived from transplantation of the RAS-driven GBM cells resulted in animal death within 25 days. Compared to corresponding endpoint tumors of PDGFB-driven GBMs, RAS-driven GBMs in implanted mice appeared to have higher cell density and intense Iba1⁺ TAM infiltration (Figure 2A). To verify the mesenchymal identify of RAS-driven GBM lines, we performed transcriptional profiling of cell lines isolated from RAS-driven GBM tumors. We observed an enrichment of pathways related to epithelial mesenchymal transition and Hippo signaling, the characteristic features of mesenchymal GBM (Supplementary Figure S2C-D).

To determine the effect of CSF1R inhibition on mesenchymal-like glioma growth, we treated the RAS-driven GBM syngeneic mice with PLX3397 daily starting the day after tumor cell transplantation (Figure 2B). Surprisingly, we found that PLX3397 treatment at the early phase of tumorigenesis accelerated tumor growth (Figure 2C,D). This resulted in a significant shorter animal life span in the syngeneic RAS-driven GBM mice (Figure 2E). Tumor morphology was not altered substantially following PLX3397 treatment (Figure 2F), but PLX3397 treatment did cause a decrease in the number of Iba1[+] TAMs in RAS-driven GBM tumors (Figure 2G). This contrasts to the maintenance of these cells in PDGFB-driven GBM tumors after PLX3397 treatment (Figure 1G)[134]. In addition, cell proliferation and apoptosis assayed by phospo-H3 and cleaved caspase 3 immunostaining were not significantly changed by CSF1R inhibition in the RAS-driven GBM tumors (Figure 2H, I).

We further investigated the effect of TAM inhibition on the tumor progression of established RAS-driven GBM tumors. In contrast to increased tumor growth when PLX3397 treatment was initiated immediately after implantation, we found that PLX3397 administration in animals with established RAS-driven GBM tumors did not substantially impact on tumor progression or animal survival compared with vehicle treated mice (Figure 2J-L).

To explore whether primary resistance to CSF1R inhibition is a general characteristic of mesenchymal glioma, we examined two additional mesenchymal GBM models. In a model with tumors derived from neural progenitors overexpressing PDGFRA-D842V and dnp53 [221], PLX3397 treatment did not inhibit tumor progression in the syngeneic mice with established GBM tumors (Supplementary Figure S3). Similarly, PDX mice orthotopically transplanted with human mesenchymal GSC20 [218,219] cells also failed to show responses to PLX3397 (Supplementary Figure S4). These observations suggest that targeting of TAMs with PLX3397 does not effectively inhibit tumor growth in genetically distinct mesenchymal subtype glioma models.

Targeting intrinsic PI3K signaling does not sensitize RAS-driven gliomas to PLX3397 inhibition
Activation of PI3K signaling has been associated with resistance to PLX3397 in PDGFB-driven glioma models [135]. To determine whether inhibition of PI3K signaling sensitizes RAS-driven tumors to CSF1R inhibition, we treated RAS-driven gliomas with a brain-penetrant reversible PI3K inhibitor BKM120 [135], or the combination of BKM120 and PLX3397 starting 7 days after tumor transplantation (Figure 3A). In contrast to previous studies in the PDGFB-driven relapsed PDGFB-driven glioma model [135], individual or combination treatment did not lead to significant improvement in animal survival (Figure 3B). Tumor growth and morphology were similar among treatment groups (Figure 3C-E). Mice treated with BKM120 singly or in combination with PLX3397 showed lower phospho-AKT (p-Akt) staining (Figure 3F). Inhibition of PI3K signaling alone did not have a significant

influence on the number of Iba1$^+$ cells detected, but the combination of BKM120 with PLX3397 did decrease the number of Iba1$^+$ TAMs in tumor tissues (Figure 3G, H). Similarly, the combination treatment did not impact tumor growth in a PDGFRA/dnp53-driven mesenchymal-like glioma model [221] (Supplementary Figure S5). These observations suggest that inhibition of PI3K-mediated signaling is insufficient to sensitize RAS- or PDGFRA-driven gliomas to CSF1R inhibition.

To identify other potential tumor-intrinsic regulators of resistance to CSF1R inhibition, we performed transcriptome profiling of primary cell lines from PDGFB-driven GBM, RAS-driven GBM, and PDGFRA-driven GBM tumors [221]. As expected due to their mesenchymal identities, PDGFRA- and RAS-driven GBM cell lines showed enrichment of genes associated with epithelial-mesenchymal transition (Figure 3I), such as *Cd44, Vim, Pdgfrb, Tgfbi, Loxl1,* and *Mmp2* (Figure 3J). In addition, we found that the CSF1R inhibition-resistant GBM lines (RAS- and PDGFRA-driven GBM lines) were enriched in transcripts involved in pro-inflammatory signaling pathways such as the interferon response and TNF-mediated signaling compared to CSF1R-inhibition-sensitive PDGFB-driven GBM gliomas (Figure 3I,J). These data suggest that inflammatory factors produced by tumor cells may contribute to the resistance to CSF1R inhibition.

Monocyte-derived macrophages do not have a pro-growth effect on PDGFB- or RAS-driven tumors

Macrophage ontogeny has been proposed as an important regulator of TAM functions in GBM [117]. Monocyte-derived macrophages can infiltrate tumors and produce growth factors to drive the immune-suppressive phenotype and growth of GBMs[116,117]. The chemokine receptor CCR2 is critical for the migration of monocyte-derived macrophages into the CNS[116,117]. To determine whether infiltration of myeloid-derived cells into PDGFB-driven tumors could lead to the responsiveness to CSF1R inhibition, we examined tumor growth in *Ccr2*-knockout mice, wherein the *Ccr2* gene is disrupted by *RFP* insertion [222]. Strikingly, the loss of *Ccr2* did not affect tumor growth or animal survival when PDGFB-driven GBM cells were transplanted into *Ccr2*-knockout mice (Supplementary Figure S6A, B). There was very little co-localization between RFP reporter and Iba1, suggesting that the Iba1+ TAM compartment is not replenished by CCR2+ bone marrow-derived cells (Supplementary Figure S6C,D). Accordingly, CCR2 knockout led to only modest changes in the infiltration of Iba1+ cells (Supplementary Figure S6E). While the numbers of CCR2-RFP+ cells only modestly decreased, we detected an increase in a population of RFP$^+$ cells expressing a T cell marker CD3 (Supplementary Figure S6F, G), which is consistent with a report that CCR2 deficiency leads to an increase in CD4$^+$ T cells in glioma tissues [200]. Similarly, the growth of RAS-driven tumors was not altered in *Ccr2*-knockout mice (Supplementary Figure S6H). These observations suggest that bone marrow-derived macrophages do not influence responses to CSF1R inhibition in either PDGFB- or RAS-driven glioma models.

Single-cell RNA-seq identifies differences in TAM populations between RAS- and PDGFB-driven gliomas

To dissect the cellular compositions and functions of TAMs in RAS- and PDGFB-driven tumors, we performed 10X Chromium single-cell RNA-seq (scRNA-seq) analysis of primary RAS-driven gliomas and compared it with single cell transcriptomic data from PDGFB-driven gliomas induced by PDGFB and dnp53 [217]. Transcriptome data were analyzed from 15,565 cells from RAS-driven GBM and 16,773 cells from PDGFB-driven GBM. Unsupervised clustering analysis identified eight different clusters with distinct gene expression signatures (Figure 4A, B). The clusters included tumor cells and immune cell populations (e.g., microglia, microphages, T cells, B cells, dendritic cells, granulocytes, pericytes, and endothelial cells). The scRNA-seq analysis was also performed using the highly parallel droplet-based single-cell transcriptomics (Drop-seq) platform [223], and similar cellular clusters were identified (Supplementary Figure S7).

To assess the heterogeneity of TAMs, we identified populations of TAMs expressing pan-macrophage markers such as *C1qa*, *Csf1r*, and *Spi1* in PDGFB- and RAS-driven tumors. In PDGFB-driven tumors, 26% of the cells were TAMs, whereas in RAS-driven tumors, 5.7% of cells were TAMs among the total cell populations (Figure 4C). Clustering of TAMs based on gene expression patterns identified cellular characteristics of cells previously reported to be present in normal brain and in brain tumors [224] including border-associated macrophages (BAM) expressing *Ms4a7*, *F13a1*, *Pf4*, and *Dab2*, microglial TAMs (*Gpr34*, *SNall1*, and *Sparc*), and monocytic TAMs (*Ccr2*, *Ifitm3*, *Ly6c2*, and *Ifitm2*) (Figure 4D, E). Brain microglial cells were the predominant TAM (76% of total) in PDGFB-driven gliomas (Figure 4D, F), while the remainder were perivascular TAMs (12%) or monocyte-derived TAMs (11%). In contrast, in Ras-driven glioma, 65.5%, expressed markers of border-associated macrophages, 15.6% were brain microglial cells, and 18.9% were monocyte-derived macrophages (Figure 4E, F), suggesting a distinct TAM pattern between PDGFB- and RAS-driven tumors.

TAM characteristics differ in RAS-driven and PDGFB-driven tumors

We next compared gene expression in TAMs from RAS-driven and PDGFB-driven tumors and identified pathways that are differentially activated. Gene set enrichment analysis (GSEA) revealed enrichment of pathways including Slit-Robo signaling, oxidative phosphorylation, PDGFRB, and neuroactive signaling in TAMs from PDGFB-driven tumors (Figure 4G). In contrast, in TAMs from RAS-driven tumors enrichment in inflammatory signaling such as TNF, interferon, NF-κB, and hypoxia pathways were observed (Figure 4G, H). Strikingly, in TAMs from RAS-driven tumors, there was upregulation of transcripts involved in antigen presentation (e.g., *H2-Ab1* and *H2-Eb1* encoding MHCII proteins), inflammation (*Il1b*), angiogenesis (*Vegfa* and *Ccl8*), and immunosuppression (*Arg1* and *Cd274*, the latter encoding checkpoint protein PD-L1) (Figure 4I). The predominant TAMs in PDGFB-driven GBM were characterized by relatively high expression of brain microglia-enriched genes (*P2ry12, Sparc, Gpr34,* and *Siglech*).

To confirm the pro-inflammatory nature of TAMs from RAS-driven tumors, we performed bulk transcriptomic profiling of endpoint RAS-driven tumors treated with vehicle or PLX3397. GSEA analysis revealed that PLX3397-treated RAS-driven tumors had lower levels of transcripts related to NF-κB signaling and phagocytosis, suggesting a decrease in tumor inflammation (Figure 4J). There was a concomitant upregulation of pro-growth signaling pathways such as Hedgehog and PI3K-FGFR signaling following macrophage depletion (Figure 4J), consistent with accelerated tumor growth after CSF1R inhibition in RAS-driven tumors. PLX3397-treated RAS-driven tumors also showed an enrichment in the proneural signature of GBM, whereas tumors from vehicle-treated mice showed an enrichment of the mesenchymal GBM signature (Figure 4K). This suggests that TAM targeting resulted in a mesenchymal to proneural transition in RAS-driven tumors, consistent with TAMs as a driver of mesenchymal GBM identity[199].

TAM subtypes in murine GBMs resemble their human counterparts

To compare the compositions of TMEs in human GBM subtypes, we examined single-cell transcriptome data of GBM tumors from publicly available datasets[77,78,215]. Human TAMs segregated into three distinct clusters (Figure 5A). All TAM clusters showed expression of pan-macrophage markers *SPI1* and *CSF1R* (Figure 5A, B). One cluster included canonical microglia as indicated by expression of *TMEM119, SALL1,* and *MEF2A*. Another cluster appear to be similar to border-associated macrophages (BAMs), which are characterized by expression of *MS4A7, CD163,* and *F13A1*. The final cluster includes proliferative TAMs that express cell proliferation markers. We found that mesenchymal subtype tumors had a higher fraction of the BAM cluster cells than the other subtypes (Figure 5C) and that TAMs in mesenchymal tumors were enriched in the genes associated with inflammatory responses, hypoxia, and NF-κB signaling (Figure 5D). Human microglia TAM and BAMs showed enrichment of signatures corresponding to the PDGFB- and RAS-driven counterparts, respectively (Figure 5E), suggesting that TAM subtypes in murine GBMs resemble those of the human counterparts.

Combination of angiogenesis and CSF1R inhibition decreases cell proliferation in RAS-driven GBM

To further understand how TME cells communicate with tumor cells in PDGFB- and RAS-driven GBM, we performed receptor ligand profiling using the CellChat pipeline [214]. We found that RAS-driven tumors had more extensive TME communication and increased TAM-to-vascular signaling than PDGFB-driven GBM (Supplementary Figure S8A, B). In PDGFB-driven GBM, endothelial cells receive IGF1 signaling from TAMs and VEGF signaling from pericytes (Supplementary Figure S8C, D). In contrast, RAS-driven GBM endothelial cells receive angiogenic signals from multiple cell types including macrophages, granulocytic cells, tumor cells, and pericytes (Supplementary Figure S8C,E). Consistent with the predicted multiple overlapping angiogenic systems, RAS-driven GBMs exhibited a higher tumor vessel density when compared to PDGFB-driven GBMs (Figure 6A, B). Notably, PLX3397 treatment caused a decrease in CD31$^+$ vessels in PDGFB-driven tumors (Figure 1), whereas vascular density was not affected by PLX3397 treatment when tumors were driven by RAS (Figure 6C).

We next examined whether co-targeting of TAMs and angiogenesis would impact the phenotype and growth of RAS-driven GBM. We treated mice implanted with RAS-driven GBM cells with PLX3397, cediranib (a VEGFR2 inhibitor), or both starting 7 days after implantation. Treatment with cediranib, PLX3397, or both yielded similar survival curves in RAS-driven GBM models (Figure 6D). The combination treatment resulted in reduced tumor mass with less dense lesions, which appeared to be infiltrative (Figure 6E, F). Strikingly, we observed that vessel density was decreased in tumors from mice treated with the combination of cediranib and PLX3397, but not in mice treated with individual agents alone (Figure 6G). These results are consistent with effects of cediranib treatment observed in other mouse glioma models [225,226]. Notably, we found that combination treatment with PLX3397 and cediranib, but not either agent alone, reduced cell proliferation in RAS-driven GBM tumors as measured by staining for the proliferative marker p-H3

(Figure 6H). Thus, our data indicate that combination of PLX3397 and anti-angiogenic treatment reduces cell proliferation and tumor mass in PLX3397-resistant RAS-driven glioma.

Discussion

The growth of GBMs is controlled not only by genetic alterations in tumor cells but also by microenvironment fitness. The brain has a unique microenvironment that has evolved to dampen immunity. How genetically distinct gliomas reprogram this microenvironment to support cancer progression remains poorly understood. A better understanding of the role of the TME will facilitate the development of innovative and efficacious strategies for targeting brain tumors. Current therapies focus mostly on targeting the glioma tumor cells without accounting for TME constituents. In this study, we show that targeting unique microenvironmental TAMs present in disparate glioma subtypes can act as a double-edge sword, contributing to the suppression of cancer cells in the PDGFB-driven model but promoting their growth and survival in other glioma models. Thus, our data reveals a context-dependent role of TME on glioma growth and progression.

TAMs consist of heterogeneous cellular populations. Early characterization of glioma-associated macrophages classified these cells into tumor-inhibiting M1 and tumor-promoting M2-like phenotypes [122,227]. Recent profiling studies indicate that both human and rodent TAMs are highly heterogeneous, transcriptionally diverse populations with both pro- and anti-tumorigenic characteristics distinct from M1 or M2 macrophages [115,123,124]. CSF1R is the receptor that mediates signaling that regulates production, differentiation, and function of macrophages. CSF1R inhibition of TAMs has been proposed as a therapeutic strategy for targeting the growth of gliomas [136]. Consistent with a previous study in a preclinical model of gliomas with PDGFB expression [134], we showed that CSF1R inhibition blocked tumor growth and progression in a PDGFB-driven glioma model. However, the CSF1R inhibitor did not significantly slow the growth of mesenchymal GBM models (RAS-driven GBM and PDGFRA-driven GBM), despite a mesenchymal to proneural transition in RAS-driven tumors after TAM targeting. Phase II clinical trials of the CSF1R inhibitor PLX3397 did not shown efficacy in patients with recurrent GBM [137]. These data suggest a tumor subtype-specific response to CSF1R inhibition. Intriguingly, our PDX models with human proneural or mesenchymal GBM failed to show any growth response or survival benefit following PLX3397 treatment, suggesting that GBM phenotype alone is not sufficient to predict tumor responsiveness to CSF1R inhibition.

We identified two main functionally distinct TAM populations from bulk and single-cell transcriptomic profiles in different animal models of GBMs. TAMs in proneural-like PDGFB-driven GBM tumors are enriched in tumor-associated microglia that are critical for driving tumor growth, whereas TAMs in mesenchymal-like RAS-driven GBM tumors exhibit expression profiles characteristic of BAMs or perivascular-associated macrophages that act to suppress tumor growth. In our GBM models, the predominant TAMs correspond to CNS-resident microglia or macrophages, and loss of chemotactic receptor CCR2 critical for peripheral macrophage migration into the CNS did not alter tumor growth, suggesting a minimal impact of monocyte-derived macrophages on GBM growth in these models.

TAMs were reprogrammed by PLX3397 treatment in the PDGFB-driven GBM model, consistent with a previous report [134]. In contrast, CSF1R inhibition resulted in depletion of TAMs in the RAS-driven model. The mechanisms underlying the differential effects of CSF1R inhibition on TAM reprogramming and growth have not been defined. Nonetheless, our observations suggest a context-dependent role of TAMs in the outgrowth of distinct gliomas, indicating that the responses of GBM to CSF1R inhibitors depends on both microenvironmental TAMs and glioma subtypes.

Due to the lack of efficacy of CSF1R inhibition in GBM clinical trials, there has been a renewed focus on development of combination therapies to improve response. A previous study showed that there was activation of PI3K signaling upon tumor relapse in PDGFB-driven tumors treated with CSF1R inhibitor BLZ945 and that co-targeting of CSF1R with PI3K inhibition was sufficient to re-sensitize relapsed tumors to CSF1R inhibition [135]. However, we found that RAS-driven tumors are not responsive to inhibition of both PI3K and CSF1R signaling, suggesting that additional mechanisms or signaling pathways confer intrinsic resistance to PI3K and CSF1R inhibition. In contrast to PDGFB-driven gliomas, where paracrine secretion of IGF1 appears to be a major source of macrophage-endothelial signaling, the RAS-driven GBM tumors have high tumor vessel density likely due to a broader vascular signaling network. Co-targeting of tumor VEGF and TAMs reduced cell proliferation in RAS-driven tumors, resulting in a reduced tumor mass, although the combined treatment did not improve animal lifespan. The exact mechanism underlying the lack of enhanced survival remain unknown. Treatment-related side effects, unidentified tumor growth regulators, or increased tumor cell infiltration may counteract the reduction in cell proliferation. Nonetheless, our data suggest that combination inhibition of angiogenesis and TAMs may improve treatment response in mesenchymal-like PLX3397-resistant gliomas.

Our data indicate that CSF1R inhibition blocks the growth of PDGFB-overexpressing glioma but not growth of other genetically distinct gliomas, indicating that there are subtype- or signaling-specific responses to TAM targeting. This highlights the need to distinguish between different TME landscapes in order to use targeted therapies most effectively. In particular, we found a difference in responsiveness to CSF1R inhibition between human proneural xenografts and PDGFB-driven mouse tumors. This difference may arise either from a mismatch between human cytokines and mouse receptors, or a specific sensitive to PDGFB-driven tumor to CSF1R inhibition. PDGF mutational events such as *PDGFA* amplification or overexpression have been shown to drive initial tumorigenic events in GBM patients[228]. There are approximately 1% of human GBMs that have PDGFB overexpression. The data from our work and others suggest that proneural gliomas with PDGFB overexpression may likely be sensitive to CSF1R inhibition, whereas mesenchymal GBMs are resistant. Future efforts could broaden the spectrum of sensitive proneural tumors by reprogramming GBMs into a "PDGFB-driven" state. The presence of an inflammatory subset of TAMs in mesenchymal-like RAS-driven GBM tumors might potentially offer a therapeutic target for immunotherapy with macrophage checkpoint inhibitors such as CD47 [229] that would augment the phagocytic activity of TAMs. Combination therapy targeting different tumor compartments has the potential to lead to more robust or durable responses.

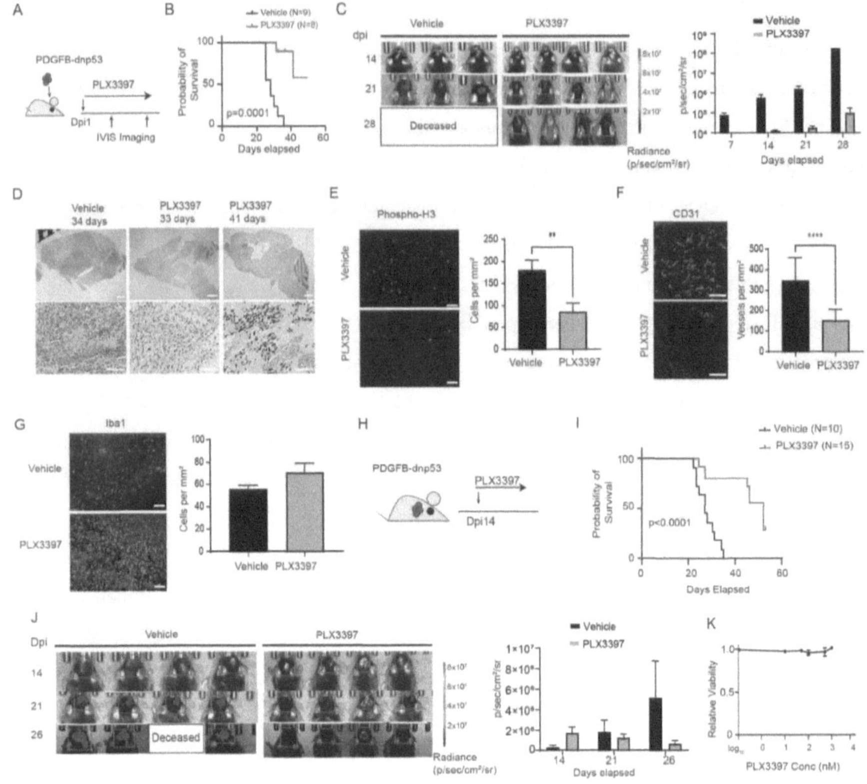

Figure 1. Targeting TAMs blocks tumor growth in murine PDGFB-driven proneural-like gliomas.

A) Experimental design of PLX3397 treatment in PDGFB-driven glioma. PLX3397 or vehicle was given every day post tumor implant (dpi). Bioluminescent imaging was performed 7, 14, 21, and 28 dpi.

B) Kaplan-Meier survival curve of mice treated with PLX3397 or vehicle from the day of implant. Significance was calculated using the log-rank test.

C) Representative bioluminescent images of mice with PDGFB-driven tumors treated with vehicle or PLX3397. Bar graph of average photon flux at each time point.

D) Representative histological images of H&E-stained sections of PDGFB-driven tumors from mice treated with vehicle or PLX3397. Scale bar (low power, 1 mm, high power,: 300 μM).

E) Left: Representative images of immunostained for p-H3 from mice with PDGFB-driven GBM treated with vehicle or PLX3397 for 3 days post tumor implant. Scale bar, 100 μM. Right: Quantification of p-H3$^+$ cells per unit area (n=3 mice, 5 fields per mouse, p=0.0028 from unpaired t-test).

F) Left: Representative images of immunostained for CD31 from mice with PDGFB-driven GBM treated with vehicle or PLX3397. Scale bar, 100 µM. Right: Quantification of CD31$^+$ cells per unit area (n=3 mice, 5 fields per mouse, p<0.0001 from unpaired t-test).
G) Representative images of immunostained for Iba1 from vehicle- or PLX3397-treated mice with PDGFB-driven GBM. Scale bar, 300 µM. Right: Quantification of Iba1$^+$ cells per unit area (n=3 mice, 5 fields per mouse, p=0.0931 from unpaired t-test).
H) Experimental design of PLX3397 treatment of mice with established PDGFB-driven tumors. Mice were treated daily with PLX3397 or vehicle beginning 14 days after tumor implantation.
I) Kaplan-Meier survival curve of mice with PDGFB-driven GBM treated with PLX3397 or vehicle daily starting at 14 days after tumor cell transplantation. P-value calculated using the log rank test.
J) Representative bioluminescent images of mice with PDGFB-driven GBM treated with vehicle or PLX3397 daily beginning 14 days post implantation. Bar graph of average photon flux at 14, 21 and 26 days.
K) Viability of 20,000 PDGFB-driven GBM cells treated with indicated concentrations of PLX3397 after 72 h of incubation relative to vehicle treated cells. Measurements were made in triplicate.

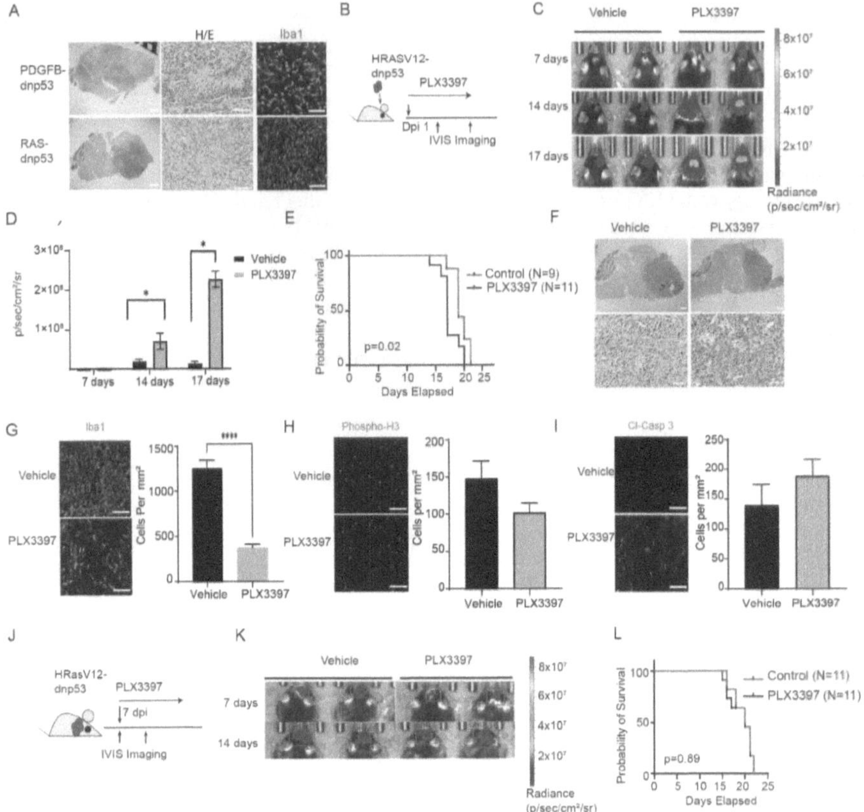

Figure 2. RAS-driven mesenchymal-like mouse gliomas are resistant to CSF1R inhibition.
A) Representative images of PDGFB- and RAS-driven tumors stained with H&E and for Iba1. Scale bar (low power, 1 mm; high power, 300 μM).
 B) Experimental design of early treatment of RAS-driven gliomas with PLX3397.
C) Representative bioluminescent images of mice with RAS-driven tumors treated with vehicle or PLX3397 beginning one day after implantation.
D) Average photon flux from RAS-driven tumors at indicated time points in mice treated with PLX3397 or vehicle (7 dpi: p=0.454, 14 dpi: p=0.036, 17 dpi: p=0.009).
E) Kaplan-Meier survival curve of mice with RAS-driven tumors treated with vehicle or PLX3397 starting one day after tumor transplantation. P-value calculated using log-rank test.
F) Representative images of H&E stained tumors from mice with RAS-driven tumors treated with vehicle or PLX3397. Scale bar (1 mm, 300 μM, and 100 μM from left to right).
G) Left: Representative images of tumors stained for Iba1 from mice with RAS-driven tumors treated with vehicle or PLX3397. Scale bar, 100 μM. Right: Quantification of Iba1$^+$ cells per unit area (n=3 mice, 3 fields per mouse, p<0.0001 by unpaired t-test).

H) Left: Representative images of tumors stained for phospho-H3 from mice with RAS-driven tumors treated with vehicle or PLX3397. Scale bar, 100 µM. Right: Quantification of phospho-H3$^+$ cells per unit area (n=3 mice, 3 fields per mouse, p=0.1260 by unpaired t-test).
I) Representative images of tumors stained for cleaved-caspase from mice with RAS-driven tumors treated with vehicle or PLX3397. Scale bar, 100 µM. Right: Quantification of cleaved-caspase per unit area (n=3 mice, 3 fields per mouse, p=0.2771 by unpaired t-test).
J) Experimental design for CSF1R inhibition in established RAS-driven tumors.
K) Representative bioluminescent images of mice RAS-driven tumors treated with vehicle or PLX3397 after tumors were established.
L) Kaplan-Meier survival curves of mice with RAS-driven tumors treated with vehicle or PLX3397 after tumors were established. P-value calculated by log-rank t-test.

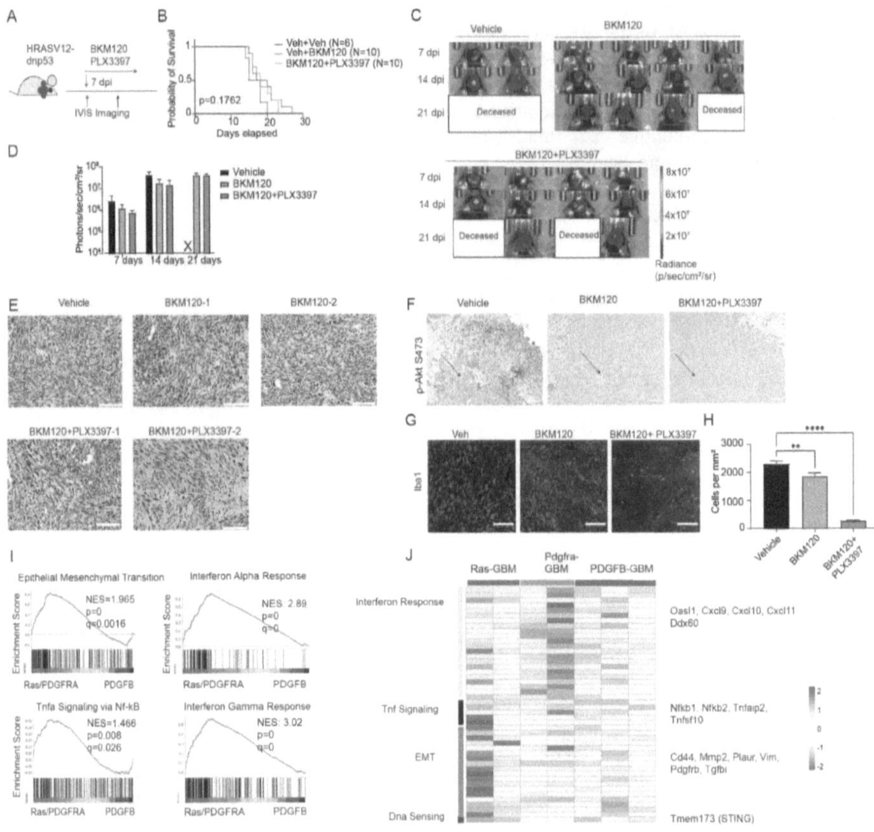

Figure 3. Tumor-intrinsic PI3K signaling does not explain resistance of RAS-driven GBM to PLX3397 inhibition.

A) Experimental design of trial of PI3K inhibitor BKM120 (20 mg/kg) in RAS-driven gliomas.

B) Kaplan-Meier survival curves of mice with RAS-driven tumors treated with PI3K inhibitor, vehicle, or combination therapy. Log rank t-test was used for significance.

C) Bioluminescent imaging of mice with RAS-driven gliomas treated with BKM120 or combination therapy.

D) Quantification of average flux from RAS-driven tumors in mice treated with vehicle, BKM120, or combination therapy. All vehicle mice deceased at 21 days.

E) Representative histology of RAS-driven tumors from mice treated with vehicle, PI3K inhibitor, or combination therapy. Scale bar, 300 μM.

F) Representative images of tumors stained for phospho-Akt S473 from mice with RAS-driven tumors treated with vehicle, BKM120, or combination therapy. Scale bar, 1 mm.

G) Representative images of tumors stained for Iba1 from mice with RAS-driven tumors treated with vehicle, BKM120, or the combination. Scale bar, 100 µM.

H) Quantification of Iba1$^+$ cells per unit area in each treatment group (n=2 mice per vehicle and BKM120, n=3 for combination, 3 fields per image). Significance with one way ANOVA + multiple comparisons (Veh vs BKM120: p=0.0074, Veh vs Combination: p<0.0001).

I) Hallmark terms enriched in transcriptomes of CSF1R resistant tumors (RAS-driven and PDGFRA-dnp53) compared to those of CSF1R-sensitive tumors (PDGFB).

J) Heatmap of genes enriched in CSF1R-resistant tumors (RAS-driven and PDGFRA-dnp53) when compared to PDGFB gliomas.

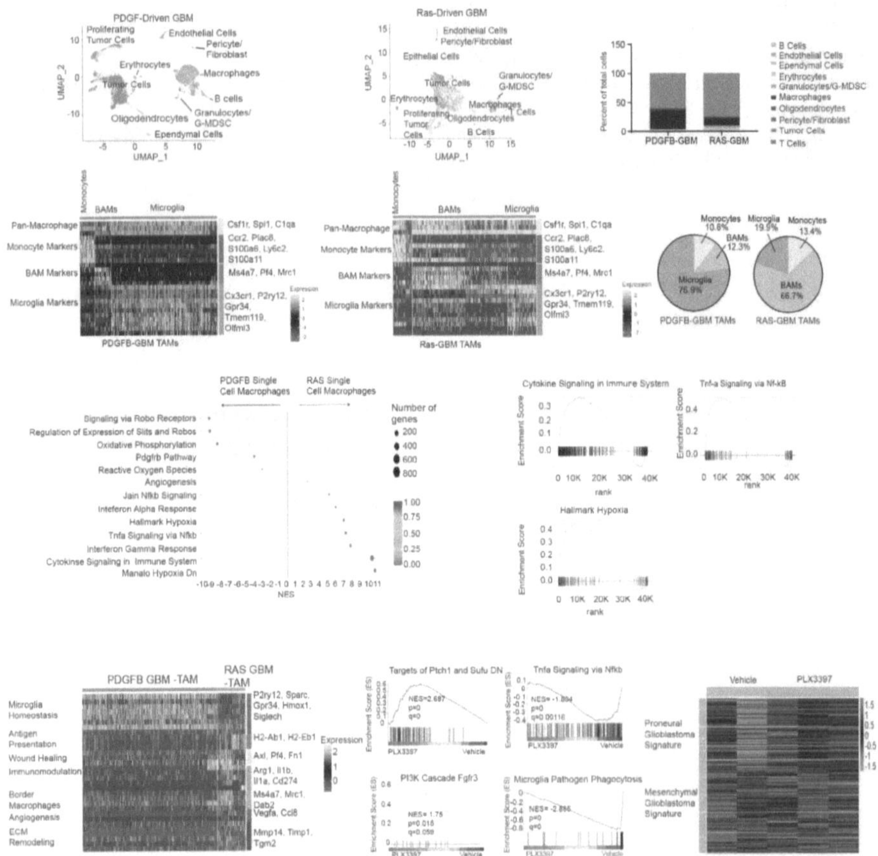

Figure 4. scRNA-seq reveals differences in TAM populations in RAS- and PDGFB-driven GBM.

A) UMAP plot of clusters derived from 10x Genomics profiling of PDGFB-driven tumors.

B) UMAP plot of clusters derived from 10x Genomics UMAP of RAS-driven glioma.

C) Bar graph of distribution of cell types in the two glioma models.

D) Heatmap of expression of marker genes for monocytes, BAMs, and microglia in PDGFB-driven gliomas.

E) Heatmap of expression of marker genes for monocytes, BAMs, and microglia in RAS-driven gliomas.

F) Pie chart showing subdivision of macrophage populations in the two GBM models.

G) Rich plot of differentially enriched terms between RAS-driven and PDGFB-driven glioma macrophages.

H) Plots of terms enriched in transcriptomes of macrophages from RAS-driven macrophages versus PDGFB-driven tumors.
I) Heatmap of differentially expressed genes in macrophages from PDGFB- and RAS-driven tumors.
J) Expression of proneural and mesenchymal signature genes from RAS-driven tumors treated with vehicle or PLX3397.

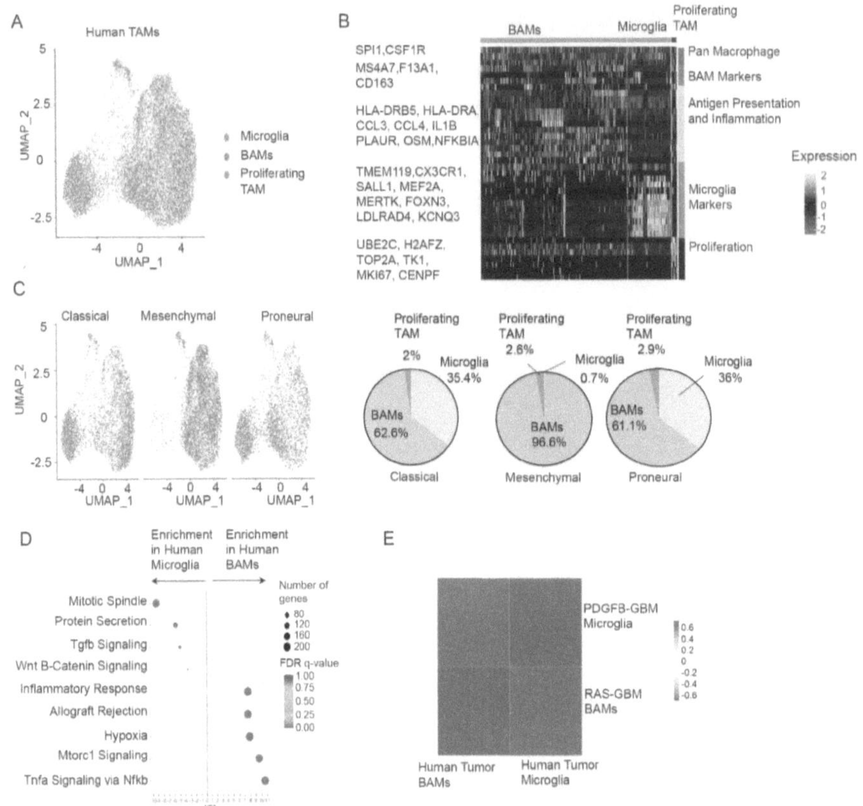

Figure 5. TAMs in human GBMs segregate based on compartment of origin.

A) UMAP plot of 10x Genomics scRNA-seq and snRNA-seq of TAMs from human GBM datasets clustered using Seuratv3 .

B) Heatmap of marker genes for BAMs, microglia, and proliferating TAMs from human GBMs. Expression of data is scaled by row.

C) UMAP plots of TAM clusters and percentages of BAMs, microglia, and proliferating TAMs in different human GBM subtypes.

D) Rich plot of Hallmark gene sets enrichment in human BAMs vs. microglia.

E) Correlation of mouse RAS-BAM and PDGFB-microglia signatures with TAMs in human GBMs

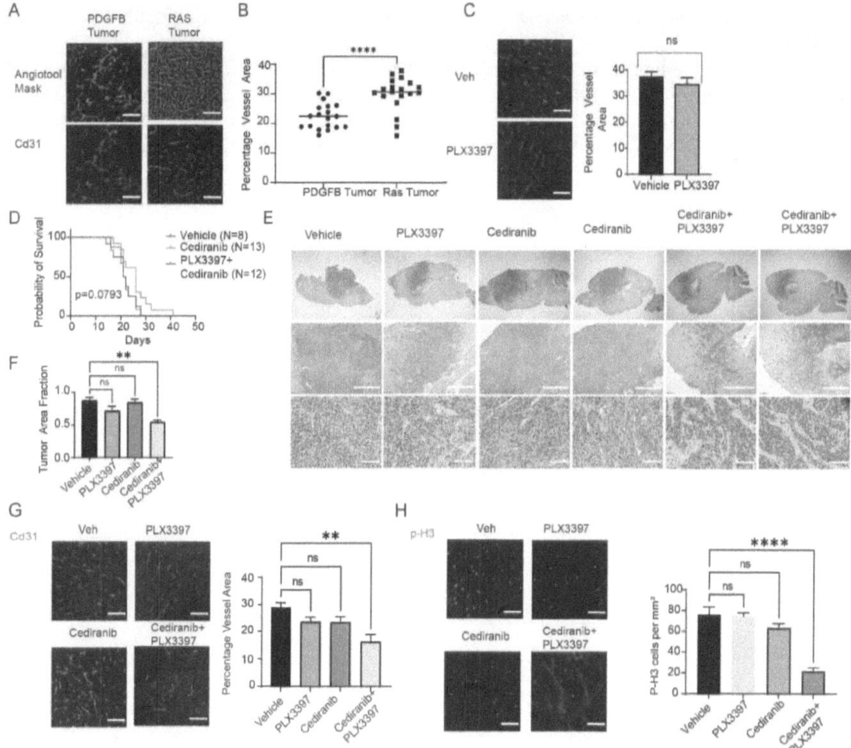

Figure 6. Combination of anti-angiogenic therapy with CSF1R inhibition decreases proliferation in RAS-driven GBM but does not improve survival outcomes.

A) Representative images of PDGFB- and RAS-driven tumors stained for Vegfa. Scale bar, 100 μM.
B) Vessel density assayed by Cd31 staining in PDGFB- and RAS-driven tumors by Angiotool (N=4, p<0.0001).
C) Representative images of CD31+ vessels from Ras-driven tumors treated with vehicle or PLX3397, scale bar, 100 μM. Quantification of percentage vessel area performed by Angiotool (right, N=3, 4 fields per sample, p=0.3083).
D) Kaplan-Meier survival curve of mice with RAS-driven tumors treated with cediranib (6 mg/kg daily), vehicle, or the combination of cediranib with PLX3397 (100 mg/kg daily).
E) Representative histology of tumors treated with vehicle, PLX3397, cediranib, or the combination therapy. Scale bar (upper/middle: 1 mm, lower: 300 μM).
F) Quantification of tumor area fraction for tumors receiving vehicle, PLX3397, cediranib or combination therapy (N=3 tumors per group). Significance calculated through one way ANOVA with multiple comparisons (PLX3397 vs Vehicle: p=0.1135, Cediranib vs Vehicle: p=0.9446, Cediranib+PLX3397 vs Vehicle: p=0.0018).
G) Left: Representative images of CD31⁺ blood vessels in endpoint tumors. Scale bar, 100 uM. Right: Quantification of vessel area (N=3, 4 fields per sample). Significance calculated using one-way

ANOVA with multiple comparisons (PLX3397 vs Vehicle: p=0.9391, Cediranib vs Vehicle: p=0.0987, Cediranib+PLX3397 vs Vehicle: p=0.0016).

H) Left: Representative images of endpoint RAS-driven tumors stained for p-H3$^+$ cells. Scale bar, 100 uM. Right: Quantification of p-H3$^+$ cells in low power fields in all treatment groups (N=3 mice per group, 4 fields per mouse). Significance calculated using one-way ANOVA with multiple comparisons (PLX3397 vs Vehicle: p=0.9944, Cediranib vs Vehicle: p=0.1784, Cediranib+PLX3397 vs Vehicle: p<0.0001).

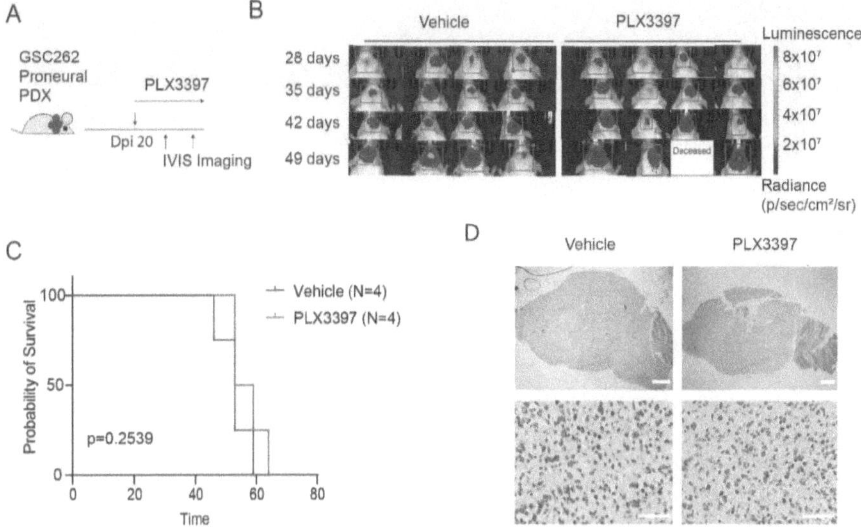

Figure S1. Human proneural PDX models do not respond to CSF1R inhibition.

A) Experimental schematic of PLX3397 treatment of GSC262 proneural PDX model.

B) Representative bioluminescent images of proneural PDX mice treated with vehicle or PLX3397 beginning 20 days post tumor implantation.

C) Kaplan-Meier survival curve of proneural PDX mice treated with vehicle or PLX3397.

D) Representative H&E-stained sections of GSC262human proneural PDX treated with vehicle or PLX3397 (scale bar: top 1mm, bottom: 300 µM).

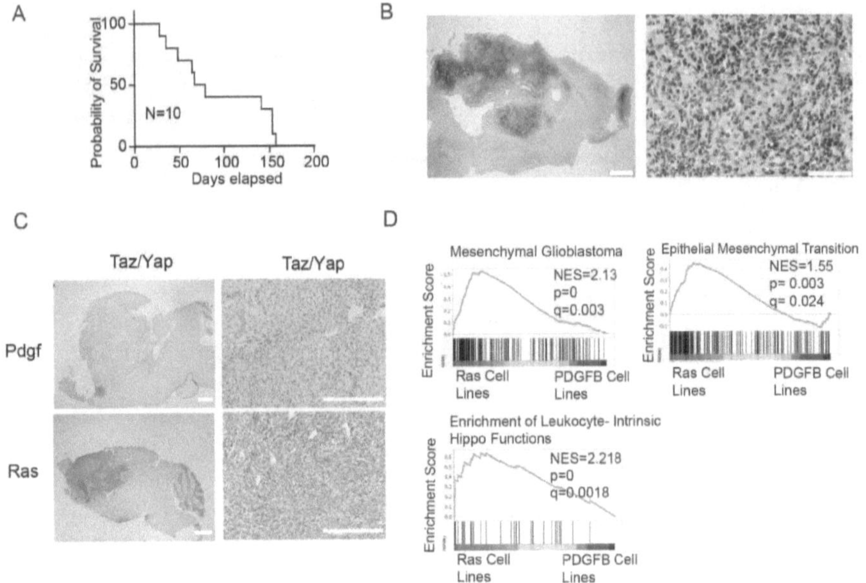

Figure S2. Tumors driven by HRASV12 and loss of p53 resemble mesenchymal GBM.

A) Kaplan Meier survival curves of adult mice injected with HRASV12-dnp53 retrovirus.

B) Representative histology of primary RAS-driven tumors.

C) Representative images of tumors stained for Yap/Taz from mice with secondary PDGFB and RAS-driven tumors (scale bar: left 1 mm, right 300 µM).

D) Terms enriched in transcriptomes of cell lines derived from RAS-driven tumors compared to PDGFB-driven tumors.

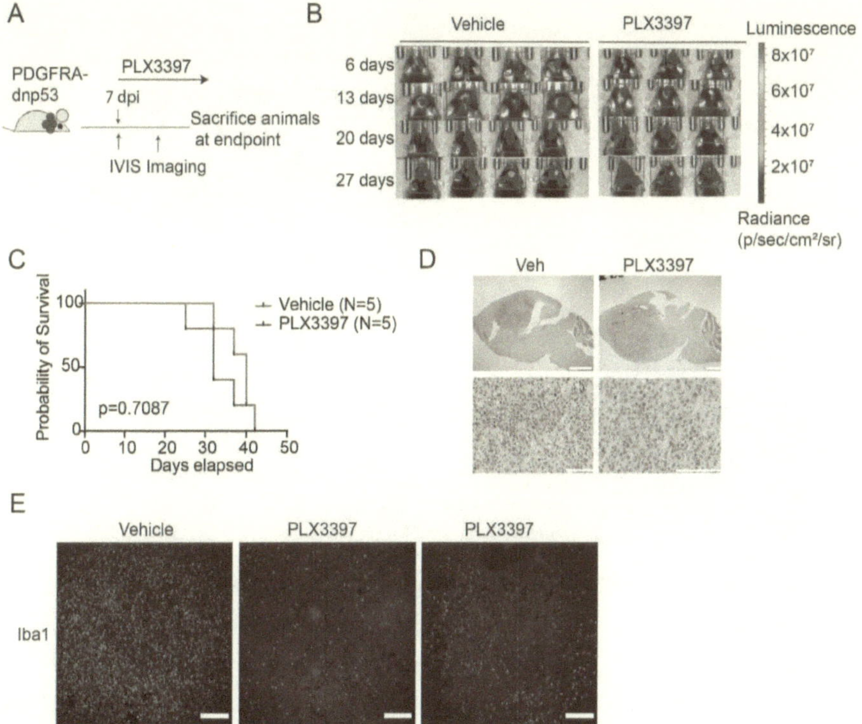

Figure S3. PDGFRA-D842V-dnp53 tumors are resistant to CSF1R inhibition.

A) Experimental design for trial of PLX3397 in PDGFRA-dnp53 tumors.

B) Kaplan-Meier survival curve of vehicle and PLX3397 treated mice with PDGFRA-dnp53 tumors.

C) Representative histology of PDGFRA-dnp53 tumors from mice treated with vehicle or PLX3397. Scale bar (top 1 mm, bottom: 300 µM)

D) Bioluminescent images of mice with PDGFRA-dnp53 tumors treated with vehicle or PLX3397.

E) Representative images of tumors stained for Iba1 from mice with PDGFRA-dnp53 tumors treated with vehicle or PLX3397. Scale bars, 200 µM.

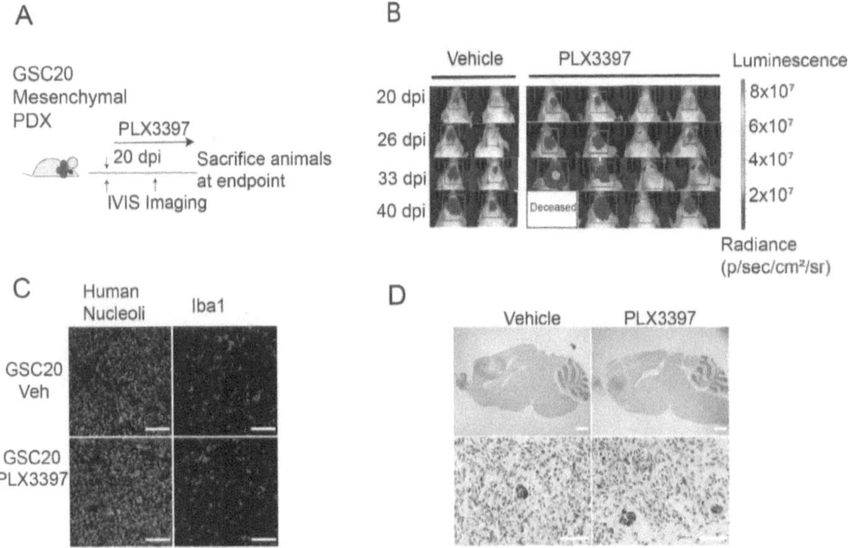

Figure S4. Mesenchymal PDX tumors are resistant to CSF1R inhibition.

A) Experimental design for trial of PLX3397 in GSC20 mesenchymal PDX tumors.

B) Bioluminescent images of mice with PDX tumors treated with vehicle or PLX3397.

C) Representative images of human nucleoli and Iba1 for GSC20 mesenchymal PDX tumors. Scale bar 100 μM.

D) Representative histology of tumors treated with vehicle or PLX3397. Scale bars, 100 μM.

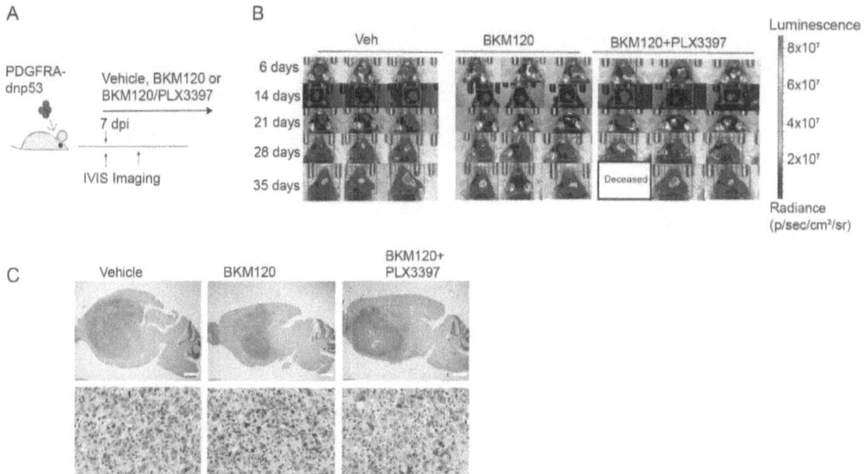

Figure S5. PDGFRA-dnp53 tumors do not respond to the combination of PI3K inhibition and CSF1R inhibition.

A) Experimental design of treatment of mice with PDGFRA-dnp53 tumors with 20 mg/kg BKM120, 100 mg/kg PLX3397, or combination therapy.

B) Representative bioluminescent images of mice with PDGFRA-dnp53 tumors from the three treatment groups.

C) Representative histology of PDGFRA-dnp53 tumors from mice treated with BKM120, PLX3397 or the combination. Scale bar (top, 1 mm, bottom, 300 μM).

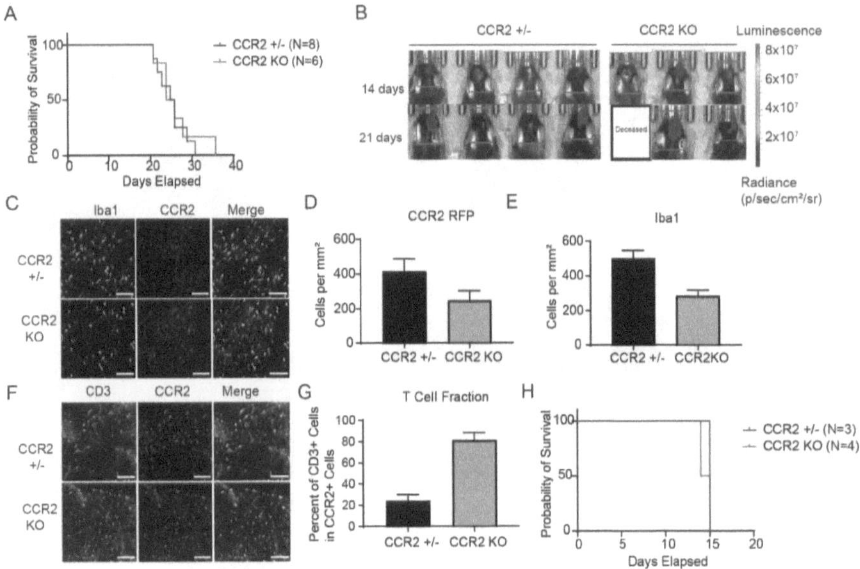

Figure S6. Bone marrow-derived macrophages do not enhance growth of syngeneic gliomas.

A) Kaplan-Meier survival curve of control ($CCR2^{+/-}$) and $CCR2^{-/-}$ mice transplanted with PDGFB-driven GBM cells.

B) Bioluminescent imaging of control and $CCR2$-knockout mice transplanted with PDGFB-driven GBM cells.

C) Representative images of tumor sections from control and $CCR2$-knockout mice stained for CCR2 and RFP. Scale bar 100 μM.

D) Quantification of the number of RFP^+ cells in PDGFB-driven tumors from control and $CCR2$-knockout mice.

E) Quantification of $Iba1^+$ cells in in PDGFB-driven tumors from control and $CCR2$-knockout mice.

F) Representative images of PDGFB-driven tumor sections from control and $CCR2$-knockout mice stained for CD3 and RFP. Scale bar 100 μM.

G) Quantification of the ratio of the number of $CD3^+$ cells to CCR2+ cells.

H) Kaplan-Meier survival curves of control and $CCR2$-knockout mice transplanted with RAS-driven tumor cells.

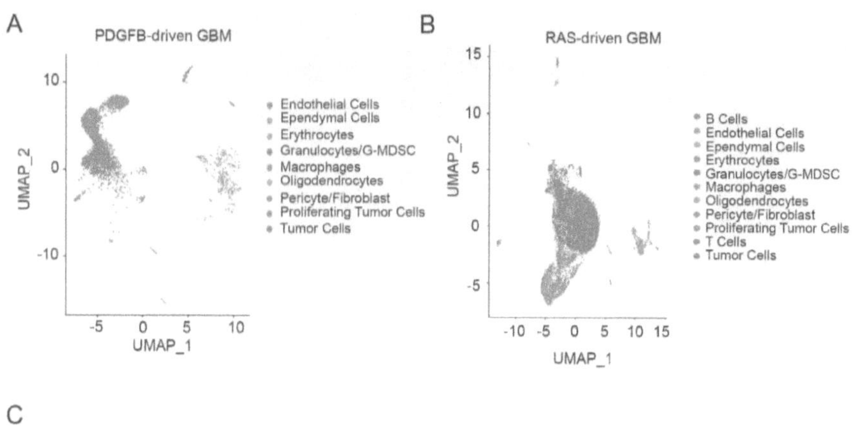

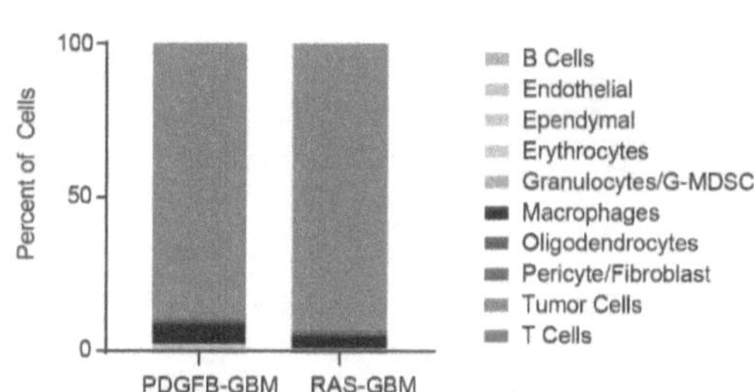

Figure S7. Drop-Seq of PDGFB- and RAS-driven tumors

Cell clustering based on A) Drop-Seq analysis of PDGFB-driven GBMs and B) RAS-driven GBMs,
C) Percentage of cells in each cluster from Drop-Seq

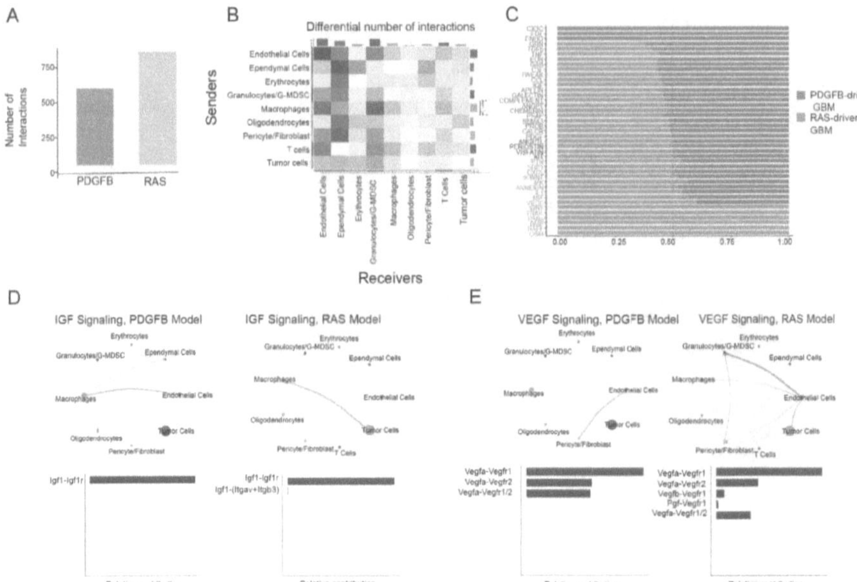

Figure S8. Differences in angiogenic signaling networks in PDGFB-driven and RAS-driven glioma.

A) Number of significant receptor-ligand interactions in PDGFB- and RAS-driven tumors.

B) Number of differential interactions between cell types in PDGFB- vs. RAS-driven tumors.

C) Differentially active receptor-ligand pathways in PDGFB- and RAS-driven tumors.

D) Igf signaling networks in PDGFB- and RAS-driven tumors.

E) Vegf signaling networks in PDGFB- and RAS-driven tumor.

Chapter 3: Proton Therapy Induces Remodeling of Tumor Cells and the Microenvironment in Glioblastoma

Introduction

Glioblastoma is a malignant brain tumor of the central nervous system with a median survival of 14 months. Radiation therapy is an important pillar of current glioblastoma treatment regimens[198]. Radiotherapy induces DNA and membrane damage through the creation of free radicals. In addition to tumor intrinsic effects, radiation therapy induces a systemic immune response that can control tumor growth at distant sites. Indeed, in some preclinical models, addition of immune checkpoint blockade to radiotherapy has shown synergistic response in vivo[142]. Despite the efficacy of radiation therapy in extending survival, gliomas inevitably recur and kill patients. Understanding the remodeling of tumors induced by radiation may be key to prolonging survival in patients.

Glioblastoma treatment often leads to long term neuro-cognitive deficits due to the effect of radiation on the developing brain[157]. Proton therapy is a method that has garnered an interest as a potential alternative to conventional x-ray based radiotherapy. Ionized protons deliver the same biologically effective dose as conventional x-rays but have a sharper drop off in radiation dose deposited after the Bragg peak when compared to conventional x-rays, allowing for increased tissue sparing. In pediatric patients, those treated with proton irradiation had better cognitive outcomes when compared to those treated with x-rays[158] and proton therapy has been found to reduce grade 3+ lymphopenia in adult patients[159]. Patients treated with proton therapy also did not see the decreases in grey and white matter volume when compared to photon treated patients[160]. Proton therapy has been deployed to treat glioblastomas therapeutically[161,162], and used palliatively to preserve quality of life in patients with recurrent glioblastomas[163]. However, the biological mechanisms underlying glioma recurrence from proton therapy remain poorly understood.

In this study, we treated an endogenous tumor model driven by PDGFB and dnp53 with several different doses of proton therapy. We found that whole brain proton therapy was able to extend survival of endogenously arising tumors, although tumors eventually relapsed. Through single cell transcriptomic profiling, we identify proton-recurrent tumors had increased infiltration of granulocytic myeloid-derived suppressor cells (G-MDSC) and macrophage reprogramming to promote tumor radioresistance. We also saw increased proneural to mesenchymal transitions in the tumor cell populations, and an expansion of tumor cells expressing oligodendrocyte-progenitor cell (OPC) signatures. Overall, we find that proton therapy is effective in treating a proneural model of mouse glioblastoma, and recurrence induces reprogramming of tumor cells and myeloid populations.

Results

Proton therapy improves survival outcomes in an endogenous model of mouse glioblastoma

To induce tumor formation, we used a previously described model that uses injection of a replication-deficient retrovirus encoding human PDGFB and dominant negative p53 into the ventricles of the neonatal brain[217]. This model induces formation of a proneural-subtype glioblastoma with characteristic features of grade IV tumors such as hemorrhage and pseudopallisading necrosis. Tumor cells resemble primitive oligodendrocyte progenitor cells (OPCs) in lineage[217], similar to diffuse intrinsic pontine gliomas and proneural adult glioblastomas [230]. We asked whether proton therapy would be effective in treating these tumors. Tumors were allowed to grow until day 37 and then were treated with 10 Gy proton irradiation delivered to the whole brain (Figure 1A). This time point was chosen because several of the mice had detectable tumors via MRI. We found that mice treated with 10 Gy showed a significantly improved survival when compared to untreated controls (Figure 1B). This is consistent with previous findings showing that both single dose and fractionated regimens of photon therapy can extend survival in PDGFB-driven glioma[118,231]. MRI images of one of the treated tumors showed a progressive increase in tumor size following radiation (Figure 1C). We found that recurrent tumors show no changes in histology (Figure 1D). We also tested two syngeneic models, the GL261 cell line transplant model and the PDGFRA-D842V-dnp53 transplant model, to examine the effects of proton therapy on other immunocompetent models of glioma. We found that similar to our endogenous model, whole brain proton irradiation was sufficient to extend survival in GL261, with no major changes in histology (Figure S1A,B). The PDGFRA-D842V-dnp53 model also displayed a dose-dependent increase in survival (Figure S1C, D). Recurrent tumors showed an increase in proliferation, but no change in stem cell marker Sox2 or astrocytic marker Gfap (Figure 2A). Recurrent tumors also did not shown any significant difference in the percentage of apoptotic cells (Figure 2C).

Recurrent tumors show increased infiltration of immunosuppressive G-MDSC

We next asked how relapse from proton therapy remodeled tumor cells and the tumor microenvironment. To address this question, we performed single cell RNA-sequencing on two independent recurrent mouse tumors after proton therapy. After filtering, we obtained 4918 cells from Drop-Seq and 14,330 cells from 10X Genomics profiling. We compared this new data to a previously published dataset from our lab of 11,836 cells gathered via 10x Genomics profiling[217]. We found that the recurrent tumor microenvironment had a decreased fraction of tumor-associated macrophages and an increased fraction of granulocytic myeloid derived suppressor cells (G-MDSC) (Figure 3A-C). Infiltration of T cells was low in both sham andrecurrent tumors , consistent with glioblastoma being an immunosuppressive tumor (Figure 3C). We next compared MDSCs from recurrent and untreated tumors. G-MDSC from sham tumors showed higher expression of glial genes *Mbp* and *Gfap*, suggesting an increased level of phagocytosis (Figure 3D). In contrast, G-MDSC from recurrent tumors had increased expression of gene sets associated with immunosuppression, such as MDSC marker S100a9, degradation of nitric oxide synthase substrate arginine(*Arg1, Arg2*), immune checkpoint expression (*Cd274, Lgals9),* adenosine production (*Entpd1, Nt5e),* cysteine sequestration (*Slc7a11, Slc3a2),* prostaglandin signaling (*Ptgs2)*, and *Tgfb1* (Figure 3E). G-MDSC also showed increased expression of genes associated with angiogenesis (*Mmp9, Prok2, Vegfa).*

To further explore how G-MDSCs operate within the ecosystem of recurrent tumors, we performed receptor-ligand profiling using the CellChat package. We found that G-MDSCs signaled to the tumor vasculature to drive angiogenesis through Vegfa, pleiotrophin[232], and oncostatin M[233] (Figure 3E, S1A). They also secreted increased macrophage chemokines such as *Csf1, Ccl6, Ccl4 and Ccl3*, suggesting they may stimulate macrophage recruitment into the tumor microenvironment (Figure 3F, S1A). Intriguingly, G-MDSC Galectin-9 interacted with checkpoint receptor Tim-3 on macrophages (Figure S2A). Tim3 signaling on macrophages has been previously reported as a negative regulator of inflammation, suggesting that G-MDSCs may act to suppress inflammatory responses on tumor-associated macrophages following radiation[234].

Recurrent TAMs are reprogrammed to promote vessel and tumor growth

We next asked what changes occurred in the tumor-associated macrophage population in relapsed tumors. Tumor-associated macrophages have been identified as important drivers of tumor growth in proneural glioma models[134,135], and also in recurrence following radiation and chemotherapy[118,128,136]. We found that tumor associated macrophages from recurrent tumors segregated transcriptionally from untreated macrophages (Figure 4A). Comparison of recurrent tumor TAMs and untreated TAMs showed upregulation of immunosuppressive molecule *Arg1*, ECM molecules *Fn1 and Tgfbi*, hypoxia markers *Ldha, Pgk1, Tpi1* and proliferative markers *Cenpx, Mki67 and Ccnd1* (Figure 4B). Recurrent TAMs also showed increased expression of bone-marrow derived macrophage marker Ccr2, while showing downregulation of canonical microglia markers such as *P2ry12, Hexb, Tmem119* and *Fcrls* (Figure 4B). Bone marrow derived macrophages have been linked to increased tumor aggressiveness[116]. TAMs showed downregulation of genes associated with inflammation such as *Fos* and *Ccl3* (Figure 4B). We next performed receptor-ligand analysis with the CellChat pipeline to demonstrate how macrophages interacted with cells in the recurrent microenvironment. Similar to the G-MDSCs, tumor-associated macrophages interacted with endothelial cells through oncostatin M and Vegf signaling (Figure S1B), two pathways that have been linked to angiogenesis. Tumor-associated macrophages in the relapsed tumor microenvironment also had increased Pdgfa/b-Pdgfra signaling, Pros1-Axl signaling, and Spp1-Cd44 signaling in the recurrent tumors when compared to untreated tumors (Figure 4C-F, S2B). Spp1-Cd44 interactions have previously been shown to be required for sphere formation *in vitro*[235] , tumor growth and radiation resistance[236]. Similarly, activation of Axl signaling promotes cell proliferation[237] and resistance to inhibitors of growth factor signaling[238].

These findings are suggestive that tumor associated macrophages act not only to stimulate blood vessel growth but also interact with tumor cells to promote regrowth after proton therapy.

Proton therapy induces OPC-like reprogramming of tumor cells

We next examined how recurrence from proton therapy remodeled tumor cell populations. Our previous work had identified that untreated tumors from Pdgfb-dnp53 tumors produced two subsets of tumor cells resembling intermediate glial precursors (iGC-like cells) and primitive oligodendrocyte progenitor cells which appear early in postnatal development (primitive-OPC-like cells)[217]. We found that the proportions of each of these populations were similar between recurrent and untreated tumors (Figure 5A-C). Radiation therapy has previously been associated with a proneural-to-mesenchymal transition in tumor cells[80]. To test if this happened in our recurrent tumors, we performed subtype analysis using updated gene signatures for proneural, classical and mesenchymal subtypes[57]. As a whole, tumor cells in relapsed tumors showed a decrease in mesenchymal signatures, an increase in classical signatures, and no change in proneural signatures (Figure 5D). Intriguingly, a subset of cells in relapsed tumors expressed a high mesenchymal signature (Figure 5E,F), suggesting a proneural-to-mesenchymal transition in a subpopulation of cells. We next examined changes in tumor cell identity by looking at the expression of a set of gene meta-modules defined in human glioma[77]. Cells in treatment naïve tumors were enriched in expression of genes associated with the AC (astrocytic) module (Figure 5G,I). In contrast, in recurrent tumors, the dominant metamodule was the OPC metamodule, resembling oligodendrocyte progenitor cells (Figure 5H, I). This is consistent with previous reports suggesting astrocytic cells may be more vulnerable to radiation, and that the classical subtype is selected against in recurrent tumors [57,58]. OPC-like signatures have been found to be associated with stemness[230] and tumor initiation[239], so this remodeling may be an adaptive change in response to radiation therapy.

Discussion

Glioblastoma is a malignant brain tumor of the central nervous system. Current treatment of adult and pediatric brain tumors includes a combination of radiotherapy with chemotherapy. However, tumors often recur following radiation therapy with more aggressive characteristics. Many patients who are frontline participants in clinical trials often have been exposed to multiple rounds of radiation and chemotherapy, but most mouse models deal with treatment naïve tumors. In this study, we profile changes in the relapsed tumor cells and microenvironment following proton beam therapy.

The cancer stem cell hypothesis argues that rare slowly dividing cells can act to repopulate the tumor following treatment. Previous work had demonstrated that NSC-like Nestin+ cells can act as a reservoir of cancer stem cells in mouse glioma[145]. In our study, we performed single cell RNA-seq of tumors treated with proton beam therapy after recurrence and find an increase in cells expressing OPC metamodules. OPC-like cells have been demonstrated to be the proliferative component of tumors, so the increased OPC-metamodule signature in relapsed tumors could be linked to tumor aggressiveness[239]. Recently, it has been shown that inhibition of PDGFRα, Erbb2 or Dnmt1 could target OPC-like lineages in glioma *in vitro* and *in vivo*[240]. These therapies may represent a reasonable approach for combination therapies in relapsed tumors with proton therapy.

The tumor microenvironment also plays an important role in recurrence from radiation therapy. Previous work had identified that tumor associated macrophages are critical in mediating relapse from conventional radiation therapy[118,128,136]. In our work, we find that tumor-associated macrophages may play a similar role in relapse from proton therapy, expressing proteins that stimulate blood vessel growth and promote tumor cell proliferation and stemness. Interestingly, tumor-associated macrophages express Pdgfa and Pdgfb, well known mitogens for oligodendrocyte progenitor cells, suggesting tumor reprogramming following radiation therapy may be mediated by TAMs. Our work suggests that similar to what has been seen with photon therapy, it may be effective to combine CSF1R inhibitors with proton therapy.

Granulocytic myeloid derived-suppressor cells express neutrophil markers and act to suppress immune responses by T cells. They accomplish this through arginine degradation, cysteine sequestration from the tumor microenvironment, increased production of immunosuppressive adenosine, and increased prostaglandin secretion. In other tumors, depletion of G-MDSCs using anti-Ly6G antibodies or CXCR2 inhibitors enhances response to radiation or immune stimulating therapy[153,241,242]. Recently, a paper has reported that a subset of GBMs with high granulocyte infiltration show increased expression of pro-angiogenic factors by tumor-associated macrophages[243]. In our work, we demonstrate an accumulation of G-MDSCs in the tumor microenvironment following recurrence from proton therapy. These cells express many of the classic molecules involved in immunosuppression such as arginase, PD-L1, and high adenosine signaling. While T cell infiltration is limited in control and recurrent Pdgfb-driven gliomas, these cells may use the Galectin-9 – Tim3 checkpoint to inhibit tumor macrophage activation from radiation-induced tumor necrosis. From receptor-ligand analysis, they act to drive angiogenesis and increase infiltration of tumor-associated macrophages. Targeting infiltration of these cells may therefore represent an avenue to reduce recurrence after proton therapy, by removing a negative regulator of anti-tumor inflammation.

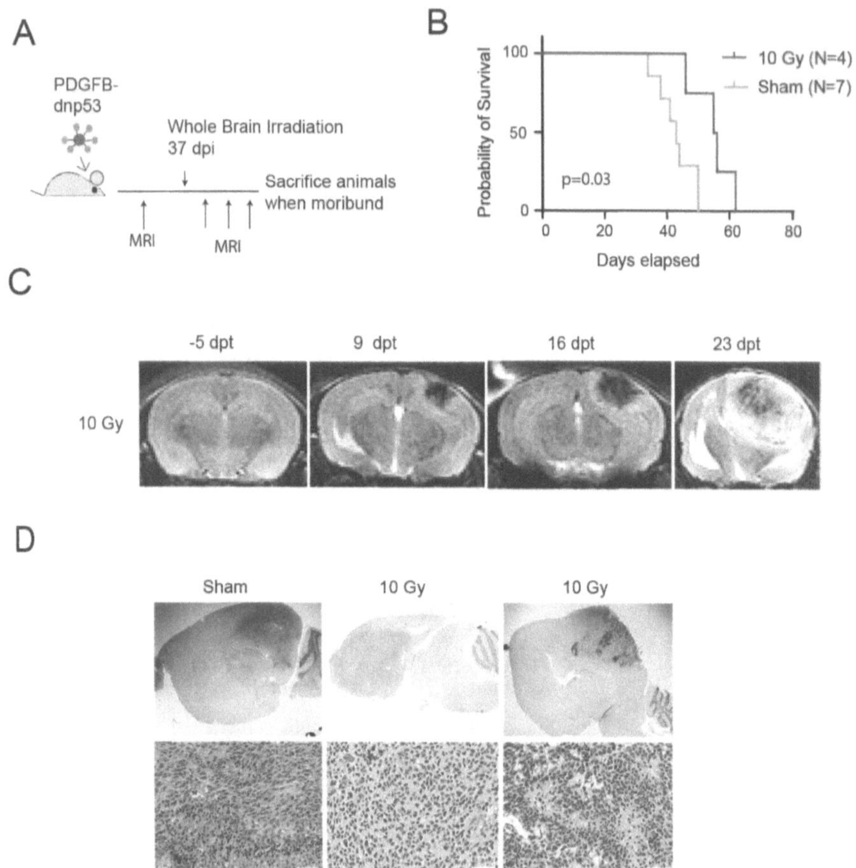

Figure 1. Proton Therapy Extends Survival of Mice with PDGFB-GBM. A) Experimental design for proton treatment of PDGFB-driven gliomas, B) Kaplan-Meier survival curve of mice treated with proton therapy. Significance was calculated using the log-rank test C) MRI images of a tumor treated with proton therapy, D) Representative histology of relapsed tumors from mice

treated with proton therapy or no treatment. Scale bar: Top: 1mm, Bottom:

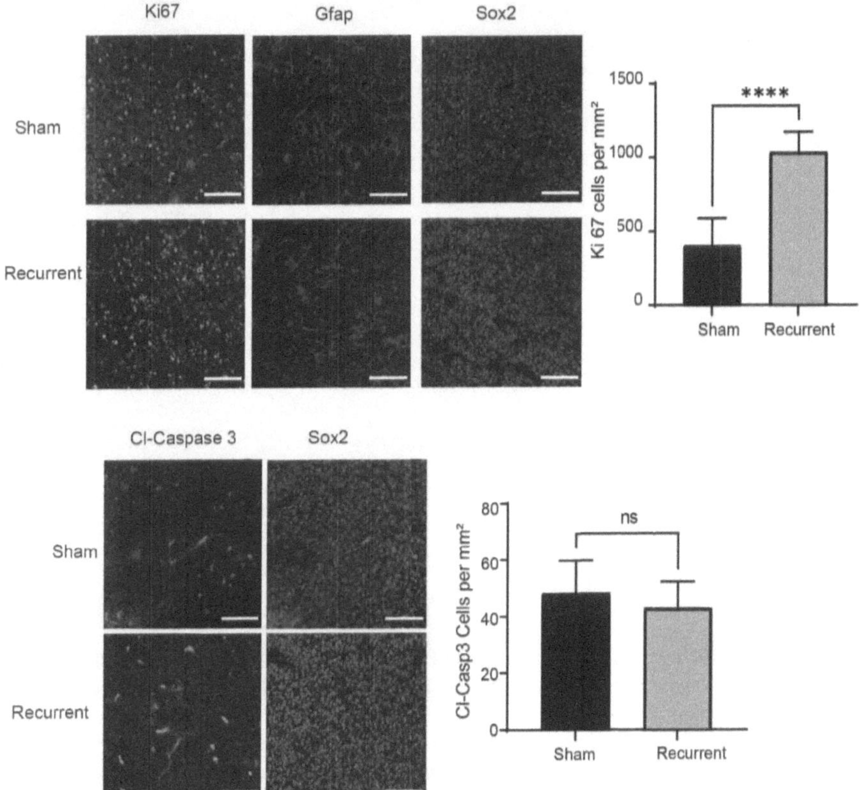

Figure 2. Proton Therapy Increases Proliferation and Does Not Affect Apoptosis. A)
Representative images of phospho-histone H3, Gfap and Sox2. Scale bar: 100 µM, B)
Quantification of phospho-H3+ cells (N=2 mice, 5 fields per mouse). C) Representative images of
cleaved-caspase 3 and phospho-H3. Scale bar: 100 µM. D) Quantification of cleaved-caspase 3
(N=2 mice, 5 fields per mouse)

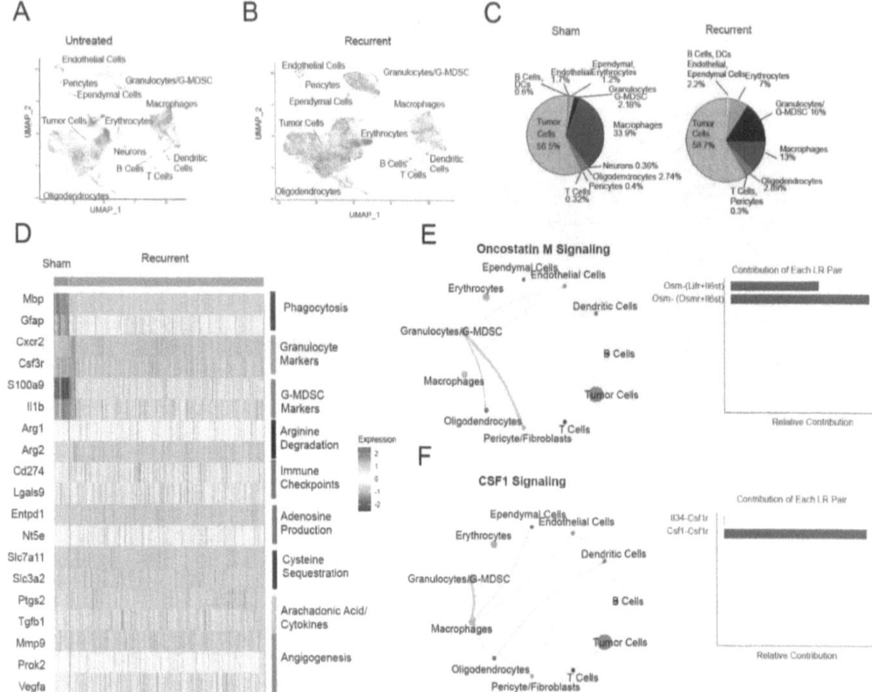

Figure 3. Relapsed Tumors Have Increased Infiltration of Immunosuppressive G-MDSCs. A) Single cell RNA-seq of sham PDGFB-dnp53 tumors (N=2), B) Single cell RNA-seq of relapsed PDGFB-dnp53 tumors after 10 Gy of proton therapy (N=2), C) Pie chart of cell composition of sham and conventional relapsed tumors, D) Heatmap of genes expressed in G-MDSCs from relapse and sham tumors, E,F) CellChat circos plot of Oncostatin M and Csf1 signaling between cell clusters in relapsed tumors

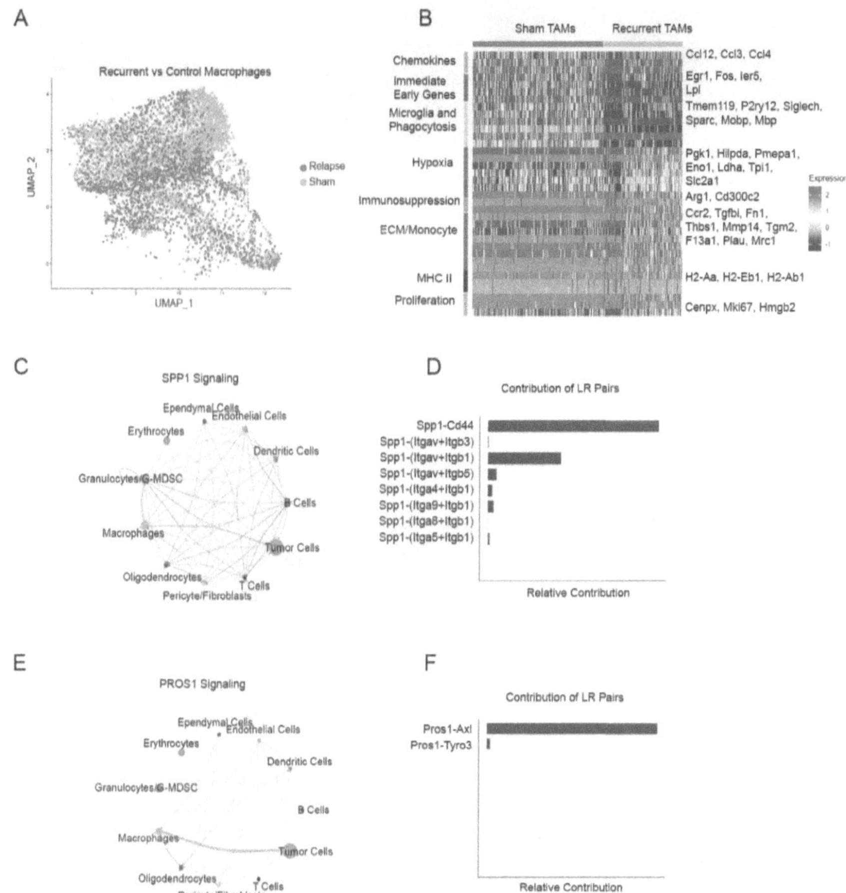

Figure 4. Proton Therapy Induces Remodeling of Tumor-Associated Macrophages. A) UMAP plot of relapsed and treatment naïve tumor-associated macrophages, B) Heatmap of marker genes of tumor associated macrophages, C,D) Analysis of receptor ligand interactions involved in Spp1 signaling in relapsed tumors, E,F) Analysis of receptor ligand interactions involved in Axl signaling in relapsed tumors

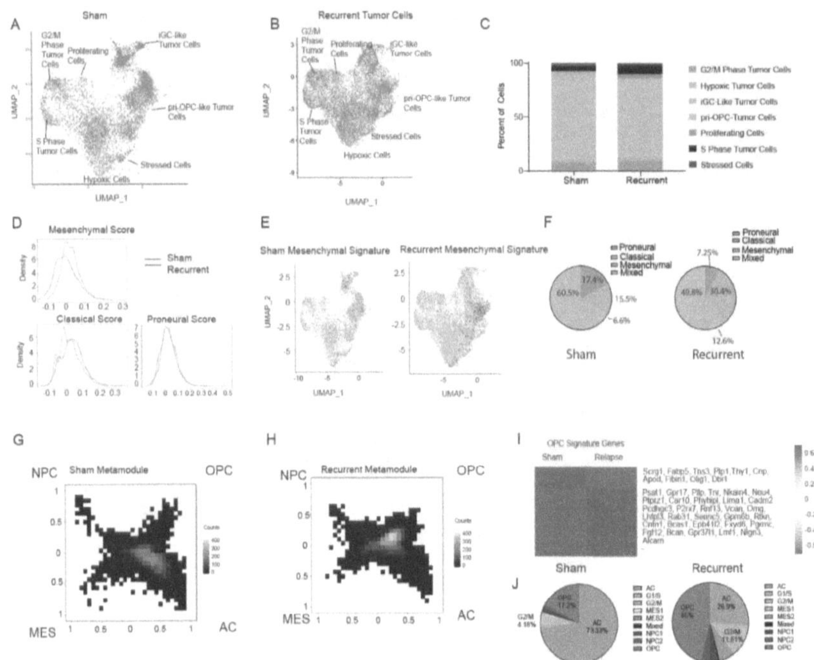

Figure 5. Proton Therapy Increases OPC Signatures in PDGFB Tumors. A) UMAP plot of tumor cell populations in treatment native tumors, B) UMAP plot of tumor cell subsets in relapsed proton tumors, C) Quantification of tumor subclusters in sham and relapsed tumors, D) UMAP plot of subtype signatures in treatment naïve tumors, E) UMAP plot of glioma subtype signatures in relapsed proton tumors, F) Pie chart of subtype frequencies in treatment naïve and relapsed tumors, G) Metamodule plot of treatment naïve tumors, H) Metamodule plot of relapsed tumors, I) Pie chart of dominant metamodules in each cell population

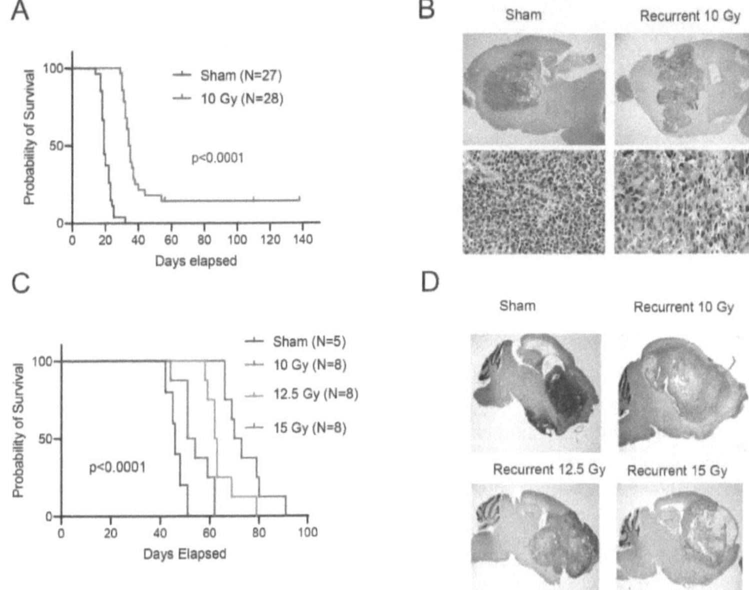

Figure S1. Proton Therapy Extends Survival in Multiple Glioma Models. A) Survival analysis of GL261 syngeneic transplant model treated with 10 Gy whole proton radiation. Statistical significance is calculated using a log rank test. B) Representative histology of mice with sham and recurrent tumors, C) Survival analysis of mice transplanted with syngeneic PDGFRA-D842V-dnp53 cells treated with varying doses of whole brain proton therapy. Significance is calculated using the log-rank test. D) Histology of mice treated with each of these doses.

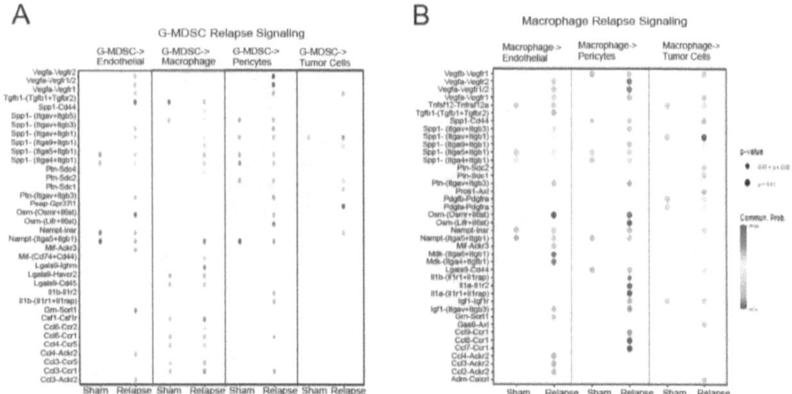

Figure S2. Proton Therapy Induces Changes in Myeloid Receptor Ligand Signaling. A) Dot plot of receptor-ligand interactions between G-MDSC and other cells in relapsed and sham tumors, B) Dot plot of receptor-ligand interactions between tumor macrophages and other cells in relapsed and sham tumors

Methods

<u>Cell Culture</u>

293T cells were cultured in DMEM + 10% FBS with 2 mM glutamine, 1 mM sodium pyruvate, 1% penicillin/streptomycin, 1X non-essential amino acids, and 10 mM HEPES. Once cells approached confluency, they were passaged using Trypsin-EDTA solution.

<u>Retroviral Packaging</u>

From a confluent dish, 293T cells were split 1:6 into new dishes. The next day, the cells were transfected using the calcium-phosphate method with 7.5 ug of transfer plasmid (pQCXIX-PDGFB-IRES-dnp53 or pQCXIX-HRasV12-IRES-dnp53) and 7.5 ug of pCL-Eco (Addgene #12371). The next day, media was changed to fresh media. Forty-eight hours after the media change, supernatants were collected, spun down at 1200 rpm to remove cellular debris and was filtered with a Corning 0.22 uM Bottle Top Vacuum Filter (Corning 430773, Corning, NY). The filtrate was then loaded into ultracentrifuge tubes (Beckman Coulter 344058, Brea, CA) and was spun at 25000 rpm in a SW28 rotor for 2 hr. Supernatant was discarded and pellet was resuspended overnight in 50 uL of HBSS at 4C. The next day, the HBSS-virus solution was aliquoted into 10 uL aliquots and kept at the -80 C until the day of injection.

<u>Animal Model of Glioma</u>

All procedures were performed in accordance with a protocol approved by the Institutional Animal Care and Use Committee at Cincinnati Children's Hospital. To generate endogenously arising tumors, P1 Balb c/J pups were anesthetized by incubation on ice for 5 min and injected with 1 uL of PDGFB-IRES-dnp53 retrovirus into the ventricles along with 2 ug/mL polybrene. After injection, pups were allowed to recover in a heated incubator and returned to their mom. Both male and female mice were used for this study.

When mice reached endpoint through development of neurological symptoms or 20% weight loss, they were anesthetized through isoflurane and perfused with cold PBS. Animals destined for histological analysis were then perfused with 4% PFA and post-fixed in PFA overnight.

<u>Proton Beam Therapy</u>

The ProBeam Pencil Beam Scanning Gantry (Varian Medical Systems, Palo Alto, CA, USA) was used to deliver a monoenergetic, single-layer transmission radiation field. Conventional doses of proton thearpy were delivered at 244 MeV. Fields were measured for flatness and symmetry using calibrated gafchromic film (Gafchromic EBT3) and calibrated flatbed scanner (Epson 10000XL). Furthermore, fields were measured for absolute dose using a calibrated Advanced Markus (PTW, Freiburg im Breisgau, Germany) ion chamber connected to a calibrated IBA Dose1 (IBA-Dosimetry, Schwarzenbruck, Germany) electrometer. The isocenter of the delivery field was determined by using diagnostic x-rays of the mouse to make sure the whole brain was irradiated. In practice, this fell at mid-ear, and was repeated consistently for multiple mice.

Mice were anesthetized using a Somnosuite Low-Flow anesthesia machine (Kent Scientific, Torrington, CT) under 2% isoflurane. Once the mice were under, they were transferred to a 3D printed holder with a nose mask to deliver continuous anesthesia, and had their paws secured by pieces of tape. Mice then received whole brain irradiation of 10 Gy for those assigned to the treatment group. After radiation, mice were transferred back to their home cage and were allowed to recover.

<u>MRI Imaging</u>

Mice were scanned using a Bruker 7T Biospec horizontal MRI system (Bruker, Billerica, MA) with a 38 mm linear coil (Bruker, Billerica, MA). Mice were anesthetized with isoflurane and kept warm with circulating air. Temperature and respiration rate were monitored with equipment from Small Animal Instruments, Inc. (SAI, Inc., NY). The mouse brain was positioned at isocenter. After acquiring localizer images, T2-weighted anatomical coronal images of the brain were acquired with a fat suppressed 2D rapid acquisition with relaxation enhancement (RARE) sequence[244] using the following parameters: TR 4 sec, TE 63 ms, echo spacing 9 ms, RARE factor 16, 17 slices, slice thickness/gap 0.8/0.2 mm, receiver bandwidth 65k, averages 5, matrix 192x192, FOV 28.4x28.4 mm, and total scan time 3:20 minutes.

<u>Hematoxylin and Eosin Staining</u>

Tissue samples were embedded in paraffin blocks by the Research Pathology Core at Cincinnati Children's Hospital and cut into 5 μM sections using a Leica microtome. Slides were deparaffinized through sequential xylene, 100% ethanol, 95% ethanol and water washes. They were incubated with Gill's Hematoxylin for 5 min, followed by Bluing Agent and a short incubation in eosin. Slides were then dehydrated and then mounted with Cytoseal (Richard Allen Scientic).

<u>Immunofluorescent Staining</u>

Sagittally cut brains were submitted to the CCHMC Pathology Core for embedding in paraffin. Once embedded, samples were cut into 5 uM sections with a Leica RM2235 microtome. Prior to staining, slides were placed in a 65 C incubator overnight to melt off excess wax. Samples were then deparaffinized through sequential changes of xylenes, 100%, 95% and 75% ethanol and moved to PBS. Samples were then put in antigen retrieval solution (10 mM sodium citrate/citric acid buffer, pH6.0) in a steamer for 35 min. Solutions were allowed to cool down to room temperature and then were washed with PBS for 5 min. They then were blocked for 2 hours in 5% NDS, 0.1% Triton-X. Primary antibodies added in 5% NDS/PBS were incubated overnight. The next day, slides were washed 5X5min in PBS and fluorescent secondary antibodies were added at a dilution of 1:500. Secondary antibodies were incubated for 1 hr, followed by counterstaining with 1 ug/mL DAPI. After another series of washes in PBS 5x5 min, the slides were mounted with Fluoromount-G and stored under dark until imaging.

Antibodies

We used the primary antibodies Rabbit anti-phospho-H3 S10 (Cell Signaling Technology, 9701S) 1:300, Rabbit anti-Cleaved Caspase 3 Asp175 (Cell Signaling Technology, 9661S, 1:200), Rat anti-mouse CXCR2 (R&D Systems MAB2164-100,1:100), Mouse anti-GFAP(1:500, Sigma G3893) and Goat anti-Sox2 1:100 (Santa Cruz, sc-17320).

For secondary antibodies, we used Donkey anti-Rabbit Alexa 594 (Jackson Immunoresearch,711-585-712), Donkey anti-Goat Alexa 488 (Jackson Immunoresearch, 705-545-147), Donkey anti-Rat Cy5 (Jackson Immunoresearch, 712-175-150), and Donkey anti-Mouse Cy5 (Jackson Immunoresearch, 715-175-151).

Dissociation of Tissues for Single-Cell RNA-seq and Sequencing of Libraries

Primary mouse tumors were dissected out of the mouse brain and minced finely with a pair of safety blades. The cell suspension was then added to a solution of TrypLE/collagenase I and incubated at 37 C for 20 min with mild agitation. Enzymes were neutralized following the addition of serum free media, and cell suspension was passed through a 40 uM filter. Cells were pelleted and resuspended in RBC Lysis Buffer (Sigma, 11814389001). After one minute of incubation, buffer was neutralized using PBS and cells were pelleted again using centrifugation. Cells were resuspended in 0.01% BSA and were submitted to the CCHMC Gene Expression Core for either Drop-Seq or 10x Chromium 3v3

Profiling. Libraries were then submitted to Novogene for sequencing. Gene barcode matrix files were generated from fastq files using the cellranger count method.

<u>Visualization and Analysis of Single-Cell RNA-seq Data</u>

Cells were analyzed using and visualized using the Seuratv3 package in R. We filtered low quality cells out, which we defined as those having less than 200 or more than 50,000 reads, or a mitochondrial percentage greater than 50%. We used the harmony package[211] to normalize batch effects between different datasets, treating each sample as its own batch. Based on the ElbowPlot function, we chose around 43 principal components for UMAP driven visualizations. Clustering was performed using the FindNeighbors and FindClusters functions in Seurat. Additionally, if the UMAP visualization showed a cluster of cells not captured by standard methods, the CellSelector command was used to define additional clusters. Markers for each cluster were defined from a combination of literature knowledge and the FindMarkers function in Seurat. Differential expression analysis was also performed using the FindMarkers function. Heatmap visualization was performed using the DoHeatmap function in R.

We used the CellChat package to perform receptor-ligand analysis on our tumors.[214] For identifying significantly expressed genes in each cluster, we used the truncatedMean option, choosing genes that were expressed in at least 10% of the cells, and told the program to consider population size in its determination of ligand importance.

<u>Calculation of Glioma Subtype Scores</u>

Gene signatures for proneural, classical and mesenchymal glioma were downloaded from the recent Wang et al publication[57]. As these signatures were human, the expression matrix genes were converted to their human analogues using the biomaRt package. Subtype scores were then calculated in a similar method as previously described[73], subtracting the mean expression of genes in the subtype from the mean expression of genes in the cell. A significance cutoff was defined by averaging all of the cells expression for each gene and then generating 4000 gene sets of size 50 to generate the random score distribution. A 5% cutoff was chosen for the score using the quantile function. Cells were assigned to a subtype based on the maximum significant score. Subtype score assignments were plotted using Seuratv3.

<u>Calculation and Visualization of Glioma Metamodules</u>

Metamodule signatures were downloaded from the Neftel et al publication[77]. Scores for each metamodule were calculated in the same manner as those for tumor subtypes shown above. For visualization, the y-coordinate was defined as $Y= \max(OPC_{score}, NPC_{score}) - \max(AC_{score}, MES_{score})$. As described previously, the x-coordinate was defined as $\pm\log_2(|OPC\text{-}NPC|+1)$ for $Y>0$ and $\pm\log_2(|AC\text{-}MES|+1)$ for $Y<0$. A negative sign was placed in front of the value for cases were NPC>OPC or MES>AC. The results were visualized using a density plot implemented using ggplot2 with the geom_bin2d and scale_fill_continuous options.

Statistics

Significance was calculated using a two-sample t-test for all quantifications. For survival curves, significance was calculated using the log-rank test.

Discussion and Future Directions

The microenvironment consists of the local milieu consisting of surrounding cells, extracellular matrix and metabolites. Integration of external signals with internal cellular state promotes cell fate decisions that drive cell fate decisions between proliferation and differentiation. My work examines this interplay in three different contexts: normal development of Schwann cells, growth of malignant glioblastomas, and the relapse of malignant brain tumors after proton therapy.

<u>Dynamic Microenvironmental Integration by mTOR is Necessary for Appropriate Schwann Cell Development</u>

Myelination of the peripheral nervous system is critical for rapid conduction of action potentials down nerve fibers and sensorimotor integration. The myelinating cells of the peripheral nervous system, Schwann cells, develop from trunk neural crest, and differentiate based on microenvironmental cues from the extracellular matrix and their associated axons[16,172,174,175]. Understanding how Schwann cells integrate microenvironmental cues to drive appropriate differentiation may offer clues to treating dysmyelinating neuropathies. The mTORC1 signaling pathway has been shown to integrate nutrient availability and growth factor signaling to promote cell growth[28], and loss of mTOR signaling in developing Schwann cells leads to defects in myelination[37,39]. The role of mTOR hyperactivity on Schwann cell development, in contrast, is not well understood. Hyperactivation of Akt, an upstream regulator of mTOR, can stimulate hypermyelination[43,45,191], but this finding is confounded by the fact that Akt can stimulate lipid synthesis through SREBPs independent of its effect on mTOR.

We found that hyperactivation of mTOR following loss of Tsc1 led to hypomyelination in the peripheral nervous system in a mTOR-dependent manner. Hypomyelination was accompanied by an increase in Schwann cell proliferation and a block in Schwann cell differentiation at the pro-myelinating stage[184,189]. Notably, loss of one allele of mTOR did not completely rescue the differentiation arrest at the pro-myelinating stage. This may be indicative of a more graded dose-response to the levels of mTOR signaling for Schwann cell differentiation. Alternatively, this difference may hint at mTOR- independent effects of Tsc1 loss on Schwann cell differentiation. As locking Schwann cells in a mTOR-active or mTOR-inactive state impairs myelination of peripheral nerves, our data suggests that dynamic regulation of mTOR signaling is critical for Schwann cell development.

We found that mTOR has a stage specific role in regulation of myelination. In contrast to its effect on developing Schwann cells, loss of Tsc1 in mature Schwann cells can lead to the formation of redundant myelin. This finding of myelin growth following Tsc1 knockout in mature myelin cells has been independently confirmed by two other groups[184,189]. Myelin growth after early development is driven by a combination of PI3K-Akt-mTOR signaling, which drives radial growth, and Yap/Taz activation from nerve stretching, which drives longitudinal growth[196]. Our lab has shown that loss of Taz/Yap promotes hypomyelination and hyperactivation of Yap/Taz can drive excessive Schwann cell proliferation and tumorigenesis[171,213]. Future work may elaborate the role of the balance between Hippo and mTOR signaling for appropriate postnatal myelination. Our work may have implications for pathological high mTOR states, like seen in patients who suffer from overnutrition and obesity.

We found that Tsc1 iKO sciatic nerves also showed an increase in cell expressing the pro-myelin marker Oct6. Oct6 is normally not expressed in mature Schwann cells, but has been shown to be induced by nerve injury[197], where Schwann cells partially dedifferentiate in order to remyelinate regenerating peripheral nerves[12]. Reactivation of mTORC1 signaling has been shown to be necessary for remyelination of sciatic nerves after crush injury[40]. Loss of Tsc1 in mature Schwann cells may therefore inappropriately activate genes associated with remyelination after injury in the resting state. While this process is deleterious in resting nerves, activation of mTORC1 signaling may represent a strategy to improve remyelination following nerve injury.

In contrast to other work which identified direct suppression of Krox20 by mTOR signaling[189], we identified Polo-like kinase signaling as a regulator of Schwann cell myelination. Consistent with its role as a driver of Schwann cell development, Plk1 levels decrease as Schwann cells differentiate from immature to myelinating Schwann cells. Inhibition of PLK1 signaling decreases Schwann cell proliferation *in vitro* and *in vivo* and can rescue hypomyelination induced by mTOR hyperactivity (Figure 1). While other groups have noted an increase in cells expressing M-phase markers following Tsc1/2 knockout in Schwann cells[184], our work is the first to show a mechanism. Since our paper has been published, a study using Tsc2 null cells demonstrated more rapid G2/M progression following checkpoint arrest from DNA damage, mediated by increased PLK1 levels[192]. Higher levels of mTORC1 activity may plausibly be a mechanism that developing Schwann cells use to promote faster cell cycle transit times during early Schwann cell development. While decreasing cell cycle times has been shown to be a mechanism to regulate differentiation in the central nervous system[193], the same has yet to be demonstrated for myelinating glia of the peripheral nervous system. Future studies may examine how cell cycle timing evolves as Schwann cells undergo differentiation. Intriguingly, PLK1-mediated proliferation of Schwann cells is reactivated in malignant tumors derived from Schwann cells, and may represent a therapeutic vulnerability of these tumors[245].

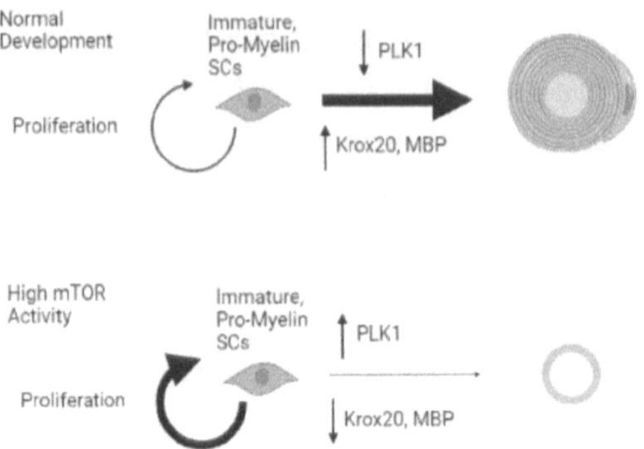

Figure 1. Hyperactivation of mTOR signaling drives hypomyelination by altering the proliferation-differentiation balance. Created in Biorender.

One important limitation of our work is that it was done in a mouse model, and therefore may not recapitulate human sciatic nerve development completely. To study the role of mTOR signaling in human Schwann cells, we will use induced pluripotent stem cells to mimic the process of Schwann cell differentiation. Protocols for human Schwann cell differentiation differentiate iPSCs to neural crest and then to Schwann cells through sequential media changes[246]. To test the effect of mTOR activity on Schwann cell development from iPSCs, we would generate iPSC lines expressing a doxycycline-inducible shRNA against Tsc1 (for mTOR hyperactivity) and a doxycycline inducible shRNA against Raptor (for knocking out mTORC1). At the final media change that induces differentiation from neural crest to Schwann cells, we would add doxycycline to the media to induce mTOR hyperactivity and knockout. After the normal differentiation period (14 days), we will assess Schwann cells for mature markers such as myelin basic protein and Krox20. As our previous findings suggested that downregulation of mTOR activity is necessary for appropriate differentiation of Schwann cells, we would also experiment with pulses of mTOR inhibition during the Schwann cell differentiation from neural crest to increase the number of mature cells. As our work also implicated PLK1 signaling in Schwann cell differentiation, we will also try different durations of PLK1 inhibition to enhance Schwann cell differentiation and yields. Overall this work will examine the role of mTOR signaling in Schwann cell differentiation in human cells, and use this signaling to enhance Schwann cell differentiation from iPSCs. This work could potentially improve Schwann cell differentiation protocols used for regenerative medicine.

<u>Glioma Immune Microenvironment and Function is Regulated by Tumor Transcriptional Subtypes</u>

Similarly to how Tsc1 null Schwann cells were locked into inappropriate responses to a changing microenvironment, malignant brain tumor cells possess mutations and epigenetic derangements that cause them to respond differently to external signals. Glioblastoma is an aggressive brain tumor that does not respond well to conventional chemotherapies and targeted inhibitors that attack tumor cells. There is growing interest in glioblastoma of targeting the non-malignant component of the tumor to reduce tumor growth and promote anti-tumor immune response, but the role of the microenvironment in glioblastoma is not well understood. The tumor microenvironment in glioblastoma consists of a variety of different cell types, including tumor-associated astrocytes, oligodendrocytes, neurons, blood vessels, lymphocytes, granulocytes, and tumor macrophages. Emerging evidence suggests a connection between tumor drivers and response to targeting of microenvironmental cells. Tumor-specific drivers influence the composition of the tumor microenvironment[57], and have been shown to have implications for immunotherapies targeting T cells[95].

For our study, we decided to investigate the function and response to CSF1R inhibition of tumor-associated macrophages, one of the most prominent immune cells in glioblastoma. CSF1R is the receptor that mediates signaling that regulates production, differentiation, and function of macrophages. CSF1R inhibition of TAMs has been proposed as a therapeutic strategy for targeting the growth of gliomas [136]. Consistent with a previous study in a preclinical model of gliomas with PDGFB expression [134], we showed that CSF1R inhibition blocked tumor growth and progression in a PDGFB-driven glioma model. However, the CSF1R inhibitor did not significantly slow the growth of mesenchymal GBM models (RAS-driven GBM and PDGFRA-driven GBM), despite a mesenchymal to proneural transition in RAS-driven tumors after TAM targeting (Figure 2). Interestingly, phase II clinical trials of the CSF1R inhibitor PLX3397 showed efficacy in only a subset of patients with recurrent GBM, hinting that a specific subset of gliomas are sensitive to this therapy[137]. PDGF mutational events such as *PDGFA* amplification or overexpression have been shown to drive initial tumorigenic events in GBM patients[228], but there are only approximately 1% of human GBMs that have PDGFB overexpression. The data from our work and others[134] suggest that proneural gliomas with PDGFB overexpression may likely be sensitive to CSF1R inhibition, whereas mesenchymal GBMs are resistant. Intriguingly, mouse tumors with autocrine activation of PDGFRA signaling were resistant to CSF1R inhibition, suggesting it is a paracrine effect that sensitizes signaling. PDGFB overexpression has been shown to increase vessel pericyte coverage and tumor growth[247]. Future efforts could broaden the spectrum of sensitive proneural tumors by reprogramming GBMs into a "PDGFB-driven" state, potentially by modulating pericyte coverage.

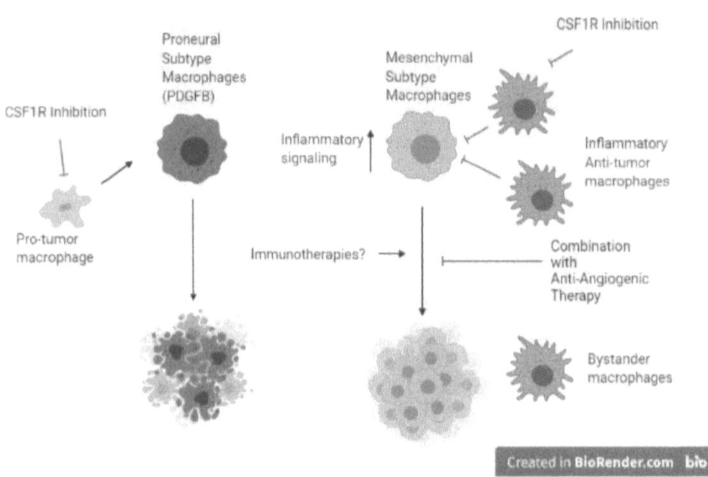

Figure 2. Distinct Subsets of TAMs in Different Glioma Subtypes

Human and rodent TAMs are highly heterogeneous, transcriptionally diverse populations with both pro- and anti-tumorigenic characteristics distinct from M1 or M2 macrophages [115,123,124]. We identified two functionally distinct TAM populations from bulk and single-cell transcriptomic profiles in different animal models of GBMs. TAMs in proneural-like PDGFB-driven GBM tumors are enriched in tumor-associated microglia that are critical for driving tumor growth, whereas TAMs in mesenchymal-like RAS-driven GBM tumors exhibit expression profiles characteristic of BAMs or perivascular-associated macrophages that act to suppress tumor growth. In our GBM models, the predominant TAMs correspond to CNS-resident microglia or macrophages, and loss of chemotactic receptor CCR2 critical for peripheral macrophage migration into the CNS did not alter tumor growth, suggesting a limited role for monocyte-derived macrophages on GBM growth in these models. The presence of an inflammatory subset of TAMs in mesenchymal-like RAS-driven GBM tumors might potentially offer a therapeutic target for immunotherapy with macrophage checkpoint inhibitors such as CD47 [229] that would augment the phagocytic activity of TAMs.

To study the potential druggability of mesenchymal TAMs more closely, we will use several different approaches. First, we will ask whether the ability of CSF1R inhibition to accelerate tumor growth is dependent on infiltration of adaptive immune cells. To this end, we will transplant Ras tumor cells into immunodeficient mice lacking T cells and NK cells and repeat our early CSF1R inhibitor experiment. If we see the same results as in an immunocompetent mouse, this suggests that mesenchymal TAMs suppress tumor growth directly. we would then test therapies that boost inflammatory signaling in macrophages such as CD40 stimulation as a mode to treat Ras driven tumors. In contrast, if there is no growth acceleration in an immunodeficient mouse, this suggests the key mode of tumor suppression is T or NK cell dependent. In this case, we would explore methods to boost T cell infiltration into tumors, such as induction of endogenous retroviruses from low dose decitabine or immunogenic chemotherapy. We will also examine transcriptional changes in tumor-associated macrophages at different stages of mesenchymal tumor growth, because our data had shown that depletion of Ras-driven macrophages at an early stage rather than a late stage, accelerated tumor growth. To explore this question, we will collect Ras macrophages at 4 days and 11 days by dissociating syngeneic Ras tumors and enriching for macrophages using Cd11b+ beads. We will perform RNA-seq of macrophages and Cd11b- tumor cells at these two time points to identify factors that are lost in macrophages when going from early stage to large tumors, and also potential macrophage suppressive molecules gained by tumor cells to promote immune evasion. I would then validate these factors through knockdown experiments performed with tumor cell lines or CX3CR1-Cre mouse lines.

One potential weakness of the models we used is that mouse tumor-microenvironment networks may not completely resemble their human counterparts. In our work, we explored this question through transplants of patient derived tumor lines into immunodeficient mice brains. However, this model may be limited as the human cytokines secreted by patient glioma cells may not be able to interact strongly with mouse receptors, leading to a different immune microenvironment. To get around this problem, we will use a recently described chimeric microglia model, in which iPS-derived hematopoetic progenitor cells are transplanted into brain of neonatal MITRG mice[248]. This model has formation of mature human microglia by two months. Following maturation of microglia, we could then transplant tumor cells from human xenografts into the mouse brain and use this system as a vehicle to study human microglia-tumor interactions in an *in vivo* setting. Furthermore, the transplantation of microglia may address an additional problem with studying the tumor microenvironment, namely, the low throughput of testing different regulators of TAM function. We could use our MITRG system to modify iHPCs before transplantation into the mice using inducible shRNAs targeting genes of interest, transplant tumor cells, and then observe the effect of knocking out genes in human TAMs on tumor growth. If interference from endogenous mouse TAMs was too much of an issue, we could cross our MITRG with FIRE mice, which lack microglia[249]. These studies would allow us to set up a human TAM- tumor cell interaction model system that would parallel our syngeneic mouse models.

Due to the lack of efficacy of CSF1R inhibition in GBM clinical trials, there has been a renewed focus on development of combination therapies to improve response. A previous study showed that there was activation of PI3K signaling upon tumor relapse in PDGFB-driven tumors treated with CSF1R inhibitor BLZ945 and that co-targeting of CSF1R with PI3K inhibition was sufficient to re-sensitize relapsed tumors to CSF1R inhibition [135]. However, we found that RAS-driven tumors are not responsive to inhibition of both PI3K and CSF1R signaling, suggesting that additional mechanisms or signaling pathways confer intrinsic resistance to PI3K and CSF1R inhibition. In contrast to PDGFB-driven gliomas, where paracrine secretion of IGF1 appears to be a major source of macrophage-endothelial signaling, the RAS-driven GBM tumors have high tumor vessel density likely due to a broader vascular signaling network. Co-targeting of tumor VEGF and TAMs reduced cell proliferation in RAS-driven tumors, resulting in a reduced tumor mass, although the combined treatment did not improve animal lifespan. The exact mechanism underlying the lack of enhanced survival remain unknown. Treatment-related side effects, unidentified tumor growth regulators, or increased tumor cell infiltration may counteract the reduction in cell proliferation. Nonetheless, our data suggest that combination inhibition of angiogenesis and TAMs may improve treatment response in mesenchymal-like PLX3397-resistant gliomas.

Our study had identified in broad strokes that mesenchymal GBM models are resistant to CSF1R inhibition while PDGFB-driven GBM was sensitive but had not really identified the specific mechanism underlying this resistance. To answer this question, we would use two approaches. First, we would test to see whether acquisition of mesenchymal identity in PDGFB-driven gliomas would make them resistant to CSF1R inhibition. To this end, we would overexpress Taz-4SA, a constitutively active Hippo transcription factor that has been shown to drive mesenchymal differentiation[70] in PDGFB-driven glioma lines. We would then transplant PDGFB-Taz tumor cells into a syngeneic mouse and treat with 100 mg/kg PLX3397 daily to observe if these cells became resistant to CSF1R inhibition. We also would use an unbiased approach to identify mediators of CSF1R resistance. We would make a PDGFB cell lines stably expressing Cas9 and transduce PDGFB-Cas9 cells with a lentiviral CRISPR library against kinases and nuclear proteins. We would transplant these induced cells into syngeneic mice and treat mouse tumors with either vehicle or PLX3397. We will collect tumors from both mouse groups and perform RNA-seq to identify gRNAs that are enriched in PLX3397 treated tumors when compared to vehicle treated tumors. Factors conveying sensitivity to PLX3397 when knocked out in cells by CRISPR, will cause expansion of that cell population, leading to increased gRNA expression. Candidate gRNAs and pathways will be validated through overexpression experiments in PDGFB cell lines and knockdown experiments in resistant mesenchymal cell lines. Overall, these experiments will help with identification of additional factors underlying CSF1R sensitivity in tumors.

<u>The Glioma Microenvironment is Altered in Recurrent Tumors After Proton Therapy</u>
The tumor microenvironment also is actively remodeled by treatment. Patient gliomas are treated with a combination of radiation, chemotherapy or anti-angiogenic therapy. Radiation therapy typically has been done with x-ray irradiation, which triggers double strand breaks in tumor cells, generates oxidative stress and activates the immune response. Proton beam therapy represents a novel approach for reducing dose deposition outside of the tumor bed, as protons show very little dose deposition past the Bragg peak. While proton therapy has been widely deployed in the clinical space, there is a dearth of information about the characteristics of tumors that recur following proton therapy, impeding development of strategies to target recurrent tumors.

Previous work involving proton therapy has shown that proton therapy is effective in controlling tumor growth of syngeneic glioma lines SMA-560 and GL261[250]. However, these models do not replicate the alterations seen in normal human glioblastoma and the GL261 model is more immunogenic than human glioblastomas usually are[251]. For our study, we used a retrovirally induced glioma model in which tumors develop after delivery of human PDGFB and dominant negative p53 to subventricular zone progenitor cells. This mimics the genetic alterations seen in pediatric and adult gliomas which have amplifications of PDGFRA and loss of p53 in tumor cells. Consistent with what had been previously observed in syngeneic transplant models, we found that whole brain proton therapy was sufficient to extend survival in endogenously arising PDGFB-driven glioma models, but eventually ended up relapsing. We found that relapsed tumors resembled their parent tumors histologically, but with an increase in cell proliferation. Our work suggests that proton therapy may be effective in prolonging survival in glioblastoma.

The cancer stem cell hypothesis states that following chemoradiotherapy, slowly dividing cancer stem cells that were resistant to the therapy repopulate the cell composition of the tumor. From our single-cell RNA-sequencing of relapsed tumors, we found that relapsed tumors following proton therapy had similar subsets of tumor cells compared to untreated tumors. This may be suggestive of either repopulation from a single progenitor or incomplete depletion of all tumor populations. Transition from a proneural to mesenchymal state has also been proposed as a mechanism for resistance to radiation therapy[66]. Following relapse from proton therapy, we found that while a subcluster of cells within the tumor demonstrated increased mesenchymal signatures, the majority of cells had gene expression resembling oligodendrocyte progenitor cells (OPCs). OPC-like states have been linked to stemness in human[230] and mouse tumors, and have been linked to relapse following chemotherapy and radiation in malignant brain tumors[239,252]. Targeting of malignant OPC stem-like states in medulloblastoma through Aurora A and Hippo inhibition has shown to promote growth arrest in medulloblastoma[252]. Inhibition of these pathways in combination with proton therapy may be an approach to prevent relapse from radiation.

In addition to tumor reprogramming following radiation, we also observed remodeling of the tumor microenvironment. Myeloid derived suppressor cells (MDSCs) are a subset of myeloid cells that act to suppress immune reactions in the context of chronic infection or cancer. MDSCs can be divided into monocytic MDSC, resembling monocytes, or PMN MDSCs, resembling neutrophils. Myeloid derived suppressor cells (MDSCs) act to suppress the immune response through increased arginase 1 expression, reactive oxygen species production, and prostaglandin production[253]. Human GBM patients show a persistent increase in blood MDSCs over time, and a poorer prognosis with increased tissue MDSC infiltration[254]. In a PDGFB-driven Nestin TvA model, targeting of MDSCs increased T cell infiltration and prolonged survival[255], and combination of MDSC targeting with immune checkpoint blockade was also able to improve survival outcomes in a murine model of glioma[200]. In our work, we find increased infiltration of granulocytic MDSCs expressing PD-L1, Arg2 and CD39 following relapse from proton therapy. In additional to their immunosuppressive role, we find these cells express molecules associated with angiogenesis and macrophage recruitment. Following proton therapy, these cells make act to put the brakes on the anti-tumor immune response induced by radiation therapy. Future studies may target this cell population following proton therapy using CXCR2 inhibitors[256], metronomic chemotherapy[257] or sunitinib[255].

Tumor-associated macrophages also are thought to play an important role in relapse of tumors from chemotherapy and conventional radiation therapy[118,129]. While we found that proton therapy does not affect the numbers of tumor-associated macrophages in our model of endogenous mouse glioma, we did observe increases in gene signatures associated with hypoxia and immunosuppression in relapsed tumors. We also saw an increase in signaling interactions between macrophages and tumor cells in relapsed tumors associated with proliferation and stemness such as Pdgf signaling, Pros1-Axl signaling and Spp1-Cd44 signaling. Our findings suggest that tumor-associated macrophages may play important roles in promoting tumor regrowth after proton therapy and may represent another target against relapsed tumors.

While our work on radiation therapy provides some important preliminary information on the effectiveness of proton therapy, more work needs to be done to validate the role of the tumor microenvironment in recurrent tumors. First, we should validate the proposed roles for G-MDSCs and tumor associated macrophages in relapse. We will test this by starting mice on CSF1R inhibitors (for tumor associated macrophages) or CXCR2 inhibitors (for G-MDSC) starting the day of proton therapy. We will follow treated mice to see if concomitant microenvironment targeting with proton therapy is sufficient to extend survival.

Radiation therapy is thought to help promote anti-tumor immune responses through induction of a systemic immune response. However, our recurrent tumors are probably too long after radiation to see how these anti-tumor immune responses evolve and are suppressed. Two potential explanations could exist for failure of radiation induced immunity. Proton therapy either does not prime T cells effectively against neoantigens, or T cells are activated by neoantigen release systemically but do not infiltrate treated tumors or are inactivated. To explore changes in T cell infiltration and activation status, we plan to collect samples from PDGFB-driven tumors for single cell analysis at 4 days after proton therapy as well as 11 days after proton therapy to observe how the microenvironment evolves following proton irradiation. We will specifically look at the level of T cell infiltration, markers of exhaustion and regulatory T cells, and immunosuppressive molecules expressed by TAMs or G-MDSC. We can validate immunosuppression of the post-proton microenvironment by using a NINJA mouse model, in which neoantigen expression can be induced by CreER[258]. We would induce neoantigen expression three days before treatment, and then treat with proton therapy. If generation of neoantigen specific T cells is the block in immune response after proton therapy, we should see clearance of the tumor. If the block is due to microenvironment mediated immunosuppression, then the tumor should still recur.

Finally, we would ask the question whether other models of glioma show similar changes after recurrence. Our work shows that in both the GL261 and PDGFRA-driven glioma models, mouse survival is extended but tumors eventually recur. To evaluate changes in the tumor microenvironment composition after recurrence, we would perform immune profiling of tumor sections using antibodies against Cxcr2 (for G-MDSC), Iba1 (total macrophages), Arg1 (for M2-like macrophages), Tnf (for M1 macrophages), Cd8 (for cytotoxic T cells), FoxP3 (for regulatory T cells), and Nkp46 (for NK cells). We will supplement this staining by performing single cell RNA-sequencing from treatment naïve and recurrent PDGFRA and GL261 tumor and ask how these tumors are similar or different to recurrent PDGFB driven tumors.

Overall, my work demonstrates the role of the microenvironment in development and tumorigenesis. I have found that overactivation of the PI3K-Akt-mTOR pathway, an integrator of growth factors and cell nutrient status, promotes context dependent responses in Schwann cells with regards to myelination. With an impaired ability to respond appropriately to changes in their microenvironment, Schwann cells show derangements in their ability to exit the cell cycle on time and produce the appropriate amount of myelin to support growing nerves. I then took this concept and applied it to malignant brain tumor cells, which have mutations and epigenetic modifications that change how they interpret normal microenvironmental signals. I found that glioblastomas with different driver mutations show differing populations of tumor associated macrophages and different responses to macrophage targeting via CSF1R inhibition. Finally, I show that the glioma microenvironment is remodeled following relapse from proton therapy. These changes in the glioma microenvironment may promote immunosuppression and escape from proton therapy.

References

1.   Brennan, K.M., Bai, Y. & Shy, M.E. Demyelinating CMT--what's known, what's new and what's in store? *Neuroscience letters* **596**, 14-26 (2015).

2.   Sanmaneechai, O., *et al.* Genotype-phenotype characteristics and baseline natural history of heritable neuropathies caused by mutations in the MPZ gene. *Brain : a journal of neurology* **138**, 3180-3192 (2015).

3.   Scherer, S.S. & Wrabetz, L. Molecular mechanisms of inherited demyelinating neuropathies. *Glia* **56**, 1578-1589 (2008).

4.   Warner, L.E., Garcia, C.A. & Lupski, J.R. Hereditary peripheral neuropathies: clinical forms, genetics, and molecular mechanisms. *Annual review of medicine* **50**, 263-275 (1999).

5.   Sabatelli, M., *et al.* Autosomal recessive hypermyelinating neuropathy. *Acta neuropathologica* **87**, 337-342 (1994).

6.   Chen, S., *et al.* Disruption of ErbB receptor signaling in adult non-myelinating Schwann cells causes progressive sensory loss. *Nature neuroscience* **6**, 1186-1193 (2003).

7.   Brifault, C., *et al.* Deletion of the Gene Encoding the NMDA Receptor GluN1 Subunit in Schwann Cells Causes Ultrastructural Changes in Remak Bundles and Hypersensitivity in Pain Processing. *The Journal of neuroscience : the official journal of the Society for Neuroscience* **40**, 9121-9136 (2020).

8.   Faroni, A., *et al.* Deletion of GABA-B receptor in Schwann cells regulates remak bundles and small nociceptive C-fibers. *Glia* **62**, 548-565 (2014).

9.   Locher, H., *et al.* Distribution and development of peripheral glial cells in the human fetal cochlea. *PloS one* **9**, e88066 (2014).

10.   Salzer, J.L. Schwann cell myelination. *Cold Spring Harbor perspectives in biology* **7**, a020529 (2015).

11.   Monk, K.R., Feltri, M.L. & Taveggia, C. New insights on Schwann cell development. *Glia* **63**, 1376-1393 (2015).

12.   Jessen, K.R., Mirsky, R. & Lloyd, A.C. Schwann Cells: Development and Role in Nerve Repair. *Cold Spring Harbor perspectives in biology* **7**, a020487 (2015).

13.   Dong, Z., *et al.* Schwann cell development in embryonic mouse nerves. *Journal of neuroscience research* **56**, 334-348 (1999).

14.   Woodhoo, A., *et al.* Notch controls embryonic Schwann cell differentiation, postnatal myelination and adult plasticity. *Nature neuroscience* **12**, 839-847 (2009).

15.   Garratt, A.N., Voiculescu, O., Topilko, P., Charnay, P. & Birchmeier, C. A dual role of erbB2 in myelination and in expansion of the schwann cell precursor pool. *The Journal of cell biology* **148**, 1035-1046 (2000).

16.   Taveggia, C., *et al.* Neuregulin-1 type III determines the ensheathment fate of axons. *Neuron* **47**, 681-694 (2005).

17.   Levi, A.D., *et al.* The influence of heregulins on human Schwann cell proliferation. *The Journal of neuroscience : the official journal of the Society for Neuroscience* **15**, 1329-1340 (1995).

18.   Ozkaynak, E., *et al.* Adam22 is a major neuronal receptor for Lgi4-mediated Schwann cell signaling. *The Journal of neuroscience : the official journal of the Society for Neuroscience* **30**, 3857-3864 (2010).

19.   Bermingham, J.R., Jr., *et al.* The claw paw mutation reveals a role for Lgi4 in peripheral nerve development. *Nature neuroscience* **9**, 76-84 (2006).

20.   Yu, W.M., Feltri, M.L., Wrabetz, L., Strickland, S. & Chen, Z.L. Schwann cell-specific ablation of laminin gamma1 causes apoptosis and prevents proliferation. *The Journal of neuroscience : the official journal of the Society for Neuroscience* **25**, 4463-4472 (2005).

21.   Poitelon, Y., *et al.* YAP and TAZ control peripheral myelination and the expression of laminin receptors in Schwann cells. *Nature neuroscience* **19**, 879-887 (2016).

22.   Petersen, S.C., *et al.* The adhesion GPCR GPR126 has distinct, domain-dependent functions in Schwann cell development mediated by interaction with laminin-211. *Neuron* **85**, 755-769 (2015).

23.   Chen, Z.L. & Strickland, S. Laminin gamma1 is critical for Schwann cell differentiation, axon myelination, and regeneration in the peripheral nerve. *The Journal of cell biology* **163**, 889-899 (2003).

24.   McKee, K.K., *et al.* Schwann cell myelination requires integration of laminin activities. *Journal of cell science* **125**, 4609-4619 (2012).

25.   Saitoh, F., Wakatsuki, S., Tokunaga, S., Fujieda, H. & Araki, T. Glutamate signals through mGluR2 to control Schwann cell differentiation and proliferation. *Scientific reports* **6**, 29856 (2016).

26.   Jha, M.K., *et al.* Monocarboxylate transporter 1 in Schwann cells contributes to maintenance of sensory nerve myelination during aging. *Glia* **68**, 161-177 (2020).

27.   Liu, G.Y. & Sabatini, D.M. mTOR at the nexus of nutrition, growth, ageing and disease. *Nature reviews. Molecular cell biology* **21**, 183-203 (2020).

28.   Laplante, M. & Sabatini, D.M. mTOR signaling in growth control and disease. *Cell* **149**, 274-293 (2012).

29.   Dibble, C.C. & Manning, B.D. Signal integration by mTORC1 coordinates nutrient input with biosynthetic output. *Nature cell biology* **15**, 555-564 (2013).

30.   Bar-Peled, L., *et al.* A Tumor suppressor complex with GAP activity for the Rag GTPases that signal amino acid sufficiency to mTORC1. *Science* **340**, 1100-1106 (2013).

31.   Wolfson, R.L., *et al.* Sestrin2 is a leucine sensor for the mTORC1 pathway. *Science* **351**, 43-48 (2016).

32.   Chantranupong, L., *et al.* The CASTOR Proteins Are Arginine Sensors for the mTORC1 Pathway. *Cell* **165**, 153-164 (2016).

33.   Gu, X., *et al.* SAMTOR is an S-adenosylmethionine sensor for the mTORC1 pathway. *Science* **358**, 813-818 (2017).

34.   Tee, A.R., Manning, B.D., Roux, P.P., Cantley, L.C. & Blenis, J. Tuberous sclerosis complex gene products, Tuberin and Hamartin, control mTOR signaling by acting as a GTPase-activating protein complex toward Rheb. *Current biology : CB* **13**, 1259-1268 (2003).

35.   Menon, S., *et al.* Spatial control of the TSC complex integrates insulin and nutrient regulation of mTORC1 at the lysosome. *Cell* **156**, 771-785 (2014).

36.   Demetriades, C., Doumpas, N. & Teleman, A.A. Regulation of TORC1 in response to amino acid starvation via lysosomal recruitment of TSC2. *Cell* **156**, 786-799 (2014).

37.   Sherman, D.L., *et al.* Arrest of myelination and reduced axon growth when Schwann cells lack mTOR. *The Journal of neuroscience : the official journal of the Society for Neuroscience* **32**, 1817-1825 (2012).

38.   Beirowski, B., *et al.* Metabolic regulator LKB1 is crucial for Schwann cell-mediated axon maintenance. *Nature neuroscience* **17**, 1351-1361 (2014).

39.   Norrmen, C., *et al.* mTORC1 controls PNS myelination along the mTORC1-RXRgamma-SREBP-lipid biosynthesis axis in Schwann cells. *Cell reports* **9**, 646-660 (2014).

40.   Norrmen, C., *et al.* mTORC1 Is Transiently Reactivated in Injured Nerves to Promote c-Jun Elevation and Schwann Cell Dedifferentiation. *The Journal of neuroscience : the official journal of the Society for Neuroscience* **38**, 4811-4828 (2018).

41.   Babetto, E., Wong, K.M. & Beirowski, B. A glycolytic shift in Schwann cells supports injured axons. *Nature neuroscience* **23**, 1215-1228 (2020).

42.   Meireles, A.M., *et al.* The Lysosomal Transcription Factor TFEB Represses Myelination Downstream of the Rag-Ragulator Complex. *Developmental cell* **47**, 319-330 e315 (2018).

43.   Goebbels, S., *et al.* Genetic disruption of Pten in a novel mouse model of tomaculous neuropathy. *EMBO molecular medicine* **4**, 486-499 (2012).

44.   Goebbels, S., *et al.* Elevated phosphatidylinositol 3,4,5-trisphosphate in glia triggers cell-autonomous membrane wrapping and myelination. *The Journal of neuroscience : the official journal of the Society for Neuroscience* **30**, 8953-8964 (2010).

45.   Domenech-Estevez, E., *et al.* Akt Regulates Axon Wrapping and Myelin Sheath Thickness in the PNS. *The Journal of neuroscience : the official journal of the Society for Neuroscience* **36**, 4506-4521 (2016).

46.   Ogata, T., *et al.* Opposing extracellular signal-regulated kinase and Akt pathways control Schwann cell myelination. *The Journal of neuroscience : the official journal of the Society for Neuroscience* **24**, 6724-6732 (2004).

47.   Jiang, M., *et al.* Regulation of PERK-eIF2alpha signalling by tuberous sclerosis complex-1 controls homoeostasis and survival of myelinating oligodendrocytes. *Nature communications* **7**, 12185 (2016).

48.   Brennan, C.W., *et al.* The somatic genomic landscape of glioblastoma. *Cell* **155**, 462-477 (2013).

49.   Wang, L.B., *et al.* Proteogenomic and metabolomic characterization of human glioblastoma. *Cancer cell* (2021).

50.   Raizer, J.J., *et al.* A phase II trial of erlotinib in patients with recurrent malignant gliomas and nonprogressive glioblastoma multiforme postradiation therapy. *Neuro-oncology* **12**, 95-103 (2010).

51.   Franceschi, E., *et al.* Gefitinib in patients with progressive high-grade gliomas: a multicentre phase II study by Gruppo Italiano Cooperativo di Neuro-Oncologia (GICNO). *British journal of cancer* **96**, 1047-1051 (2007).

52.   Chinnaiyan, P., *et al.* A randomized phase II study of everolimus in combination with chemoradiation in newly diagnosed glioblastoma: results of NRG Oncology RTOG 0913. *Neuro-oncology* **20**, 666-673 (2018).

53.   Xu, W., *et al.* Oncometabolite 2-hydroxyglutarate is a competitive inhibitor of alpha-ketoglutarate-dependent dioxygenases. *Cancer cell* **19**, 17-30 (2011).

54.   Noushmehr, H., *et al.* Identification of a CpG island methylator phenotype that defines a distinct subgroup of glioma. *Cancer cell* **17**, 510-522 (2010).

55.   Yan, H., *et al.* IDH1 and IDH2 mutations in gliomas. *The New England journal of medicine* **360**, 765-773 (2009).

56.   Ceccarelli, M., *et al.* Molecular Profiling Reveals Biologically Discrete Subsets and Pathways of Progression in Diffuse Glioma. *Cell* **164**, 550-563 (2016).

57.   Wang, Q., *et al.* Tumor Evolution of Glioma-Intrinsic Gene Expression Subtypes Associates with Immunological Changes in the Microenvironment. *Cancer cell* **32**, 42-56 e46 (2017).

58.   Verhaak, R.G., *et al.* Integrated genomic analysis identifies clinically relevant subtypes of glioblastoma characterized by abnormalities in PDGFRA, IDH1, EGFR, and NF1. *Cancer cell* **17**, 98-110 (2010).

59.   Sturm, D., *et al.* Hotspot mutations in H3F3A and IDH1 define distinct epigenetic and biological subgroups of glioblastoma. *Cancer cell* **22**, 425-437 (2012).

60.   Wu, Y., *et al.* Glioblastoma epigenome profiling identifies SOX10 as a master regulator of molecular tumour subtype. *Nature communications* **11**, 6434 (2020).

61.   Wirsching, H.G., *et al.* Bevacizumab plus hypofractionated radiotherapy versus radiotherapy alone in elderly patients with glioblastoma: the randomized, open-label, phase II ARTE trial. *Annals of oncology : official journal of the European Society for Medical Oncology* **29**, 1423-1430 (2018).

62.   Sandmann, T., *et al.* Patients With Proneural Glioblastoma May Derive Overall Survival Benefit From the Addition of Bevacizumab to First-Line Radiotherapy and Temozolomide: Retrospective Analysis of the AVAglio Trial. *Journal of clinical oncology : official journal of the American Society of Clinical Oncology* **33**, 2735-2744 (2015).

63.   Tejero, R., *et al.* Gene signatures of quiescent glioblastoma cells reveal mesenchymal shift and interactions with niche microenvironment. *EBioMedicine* **42**, 252-269 (2019).

64.   de Souza, C.F., *et al.* A Distinct DNA Methylation Shift in a Subset of Glioma CpG Island Methylator Phenotypes during Tumor Recurrence. *Cell reports* **23**, 637-651 (2018).

65.   Wood, M.D., Reis, G.F., Reuss, D.E. & Phillips, J.J. Protein Analysis of Glioblastoma Primary and Posttreatment Pairs Suggests a Mesenchymal Shift at Recurrence. *Journal of neuropathology and experimental neurology* **75**, 925-935 (2016).

66.   Bhat, K.P.L., *et al.* Mesenchymal differentiation mediated by NF-kappaB promotes radiation resistance in glioblastoma. *Cancer cell* **24**, 331-346 (2013).

67. Piao, Y., *et al.* Acquired resistance to anti-VEGF therapy in glioblastoma is associated with a mesenchymal transition. *Clinical cancer research : an official journal of the American Association for Cancer Research* **19**, 4392-4403 (2013).

68. Lee, D.W., *et al.* The NF-kappaB RelB protein is an oncogenic driver of mesenchymal glioma. *PloS one* **8**, e57489 (2013).

69. Carro, M.S., *et al.* The transcriptional network for mesenchymal transformation of brain tumours. *Nature* **463**, 318-325 (2010).

70. Bhat, K.P., *et al.* The transcriptional coactivator TAZ regulates mesenchymal differentiation in malignant glioma. *Genes & development* **25**, 2594-2609 (2011).

71. Kaffes, I., *et al.* Human Mesenchymal glioblastomas are characterized by an increased immune cell presence compared to Proneural and Classical tumors. *Oncoimmunology* **8**, e1655360 (2019).

72. Conroy, S., Kruyt, F.A.E., Wagemakers, M., Bhat, K.P.L. & den Dunnen, W.F.A. IL-8 associates with a pro-angiogenic and mesenchymal subtype in glioblastoma. *Oncotarget* **9**, 15721-15731 (2018).

73. Patel, A.P., *et al.* Single-cell RNA-seq highlights intratumoral heterogeneity in primary glioblastoma. *Science* **344**, 1396-1401 (2014).

74. Snuderl, M., *et al.* Mosaic amplification of multiple receptor tyrosine kinase genes in glioblastoma. *Cancer cell* **20**, 810-817 (2011).

75. Little, S.E., *et al.* Receptor tyrosine kinase genes amplified in glioblastoma exhibit a mutual exclusivity in variable proportions reflective of individual tumor heterogeneity. *Cancer research* **72**, 1614-1620 (2012).

76. Szerlip, N.J., *et al.* Intratumoral heterogeneity of receptor tyrosine kinases EGFR and PDGFRA amplification in glioblastoma defines subpopulations with distinct growth factor response. *Proceedings of the National Academy of Sciences of the United States of America* **109**, 3041-3046 (2012).

77. Neftel, C., *et al.* An Integrative Model of Cellular States, Plasticity, and Genetics for Glioblastoma. *Cell* **178**, 835-849 e821 (2019).

78. Wang, L., *et al.* The Phenotypes of Proliferating Glioblastoma Cells Reside on a Single Axis of Variation. *Cancer discovery* **9**, 1708-1719 (2019).

79. Lan, X., *et al.* Fate mapping of human glioblastoma reveals an invariant stem cell hierarchy. *Nature* **549**, 227-232 (2017).

80. Minata, M., *et al.* Phenotypic Plasticity of Invasive Edge Glioma Stem-like Cells in Response to Ionizing Radiation. *Cell reports* **26**, 1893-1905 e1897 (2019).

81. Liu, H., *et al.* Use of Angiotensin System Inhibitors Is Associated with Immune Activation and Longer Survival in Nonmetastatic Pancreatic Ductal Adenocarcinoma. *Clinical cancer research : an official journal of the American Association for Cancer Research* **23**, 5959-5969 (2017).

82. Robert, C., *et al.* Pembrolizumab versus Ipilimumab in Advanced Melanoma. *The New England journal of medicine* **372**, 2521-2532 (2015).

83. Reck, M., *et al.* Pembrolizumab versus Chemotherapy for PD-L1-Positive Non-Small-Cell Lung Cancer. *The New England journal of medicine* **375**, 1823-1833 (2016).

84. Henrik Heiland, D., *et al.* Tumor-associated reactive astrocytes aid the evolution of immunosuppressive environment in glioblastoma. *Nature communications* **10**, 2541 (2019).

85. Kim, J.K., *et al.* Tumoral RANKL activates astrocytes that promote glioma cell invasion through cytokine signaling. *Cancer letters* **353**, 194-200 (2014).

86. Chen, W., *et al.* Glioma cells escaped from cytotoxicity of temozolomide and vincristine by communicating with human astrocytes. *Medical oncology* **32**, 43 (2015).

87. Huang, Y., *et al.* Oligodendrocyte progenitor cells promote neovascularization in glioma by disrupting the blood-brain barrier. *Cancer research* **74**, 1011-1021 (2014).

88. Tantillo, E., *et al.* Differential roles of pyramidal and fast-spiking, GABAergic neurons in the control of glioma cell proliferation. *Neurobiology of disease* **141**, 104942 (2020).

89.   Venkatesh, H.S., *et al.* Electrical and synaptic integration of glioma into neural circuits. *Nature* **573**, 539-545 (2019).

90.   Venkatesh, H.S., *et al.* Neuronal Activity Promotes Glioma Growth through Neuroligin-3 Secretion. *Cell* **161**, 803-816 (2015).

91.   Venkatesh, H.S., *et al.* Targeting neuronal activity-regulated neuroligin-3 dependency in high-grade glioma. *Nature* **549**, 533-537 (2017).

92.   Venkataramani, V., *et al.* Glutamatergic synaptic input to glioma cells drives brain tumour progression. *Nature* **573**, 532-538 (2019).

93.   Terry, R.L., *et al.* Immune profiling of pediatric solid tumors. *The Journal of clinical investigation* **130**, 3391-3402 (2020).

94.   Omuro, A., *et al.* Nivolumab with or without ipilimumab in patients with recurrent glioblastoma: results from exploratory phase I cohorts of CheckMate 143. *Neuro Oncol* **20**, 674-686 (2018).

95.   Zhao, J., *et al.* Immune and genomic correlates of response to anti-PD-1 immunotherapy in glioblastoma. *Nature medicine* **25**, 462-469 (2019).

96.   Reardon, D.A., *et al.* Effect of Nivolumab vs Bevacizumab in Patients With Recurrent Glioblastoma: The CheckMate 143 Phase 3 Randomized Clinical Trial. *JAMA oncology* **6**, 1003-1010 (2020).

97.   Yee, P.P., *et al.* Neutrophil-induced ferroptosis promotes tumor necrosis in glioblastoma progression. *Nature communications* **11**, 5424 (2020).

98.   Bertaut, A., *et al.* Blood baseline neutrophil count predicts bevacizumab efficacy in glioblastoma. *Oncotarget* **7**, 70948-70958 (2016).

99.   Liang, J., *et al.* Neutrophils promote the malignant glioma phenotype through S100A4. *Clinical cancer research : an official journal of the American Association for Cancer Research* **20**, 187-198 (2014).

100.   Wang, P.F., *et al.* Neutrophil depletion enhances the therapeutic effect of PD-1 antibody on glioma. *Aging* **12**, 15290-15301 (2020).

101.   Dubinski, D., *et al.* CD4+ T effector memory cell dysfunction is associated with the accumulation of granulocytic myeloid-derived suppressor cells in glioblastoma patients. *Neuro-oncology* **18**, 807-818 (2016).

102.   Arvanitis, C.D., Ferraro, G.B. & Jain, R.K. The blood-brain barrier and blood-tumour barrier in brain tumours and metastases. *Nature reviews. Cancer* **20**, 26-41 (2020).

103.   Mathivet, T., *et al.* Dynamic stroma reorganization drives blood vessel dysmorphia during glioma growth. *EMBO molecular medicine* **9**, 1629-1645 (2017).

104.   Sarkaria, J.N., *et al.* Is the blood-brain barrier really disrupted in all glioblastomas? A critical assessment of existing clinical data. *Neuro-oncology* **20**, 184-191 (2018).

105.   Hambardzumyan, D. & Bergers, G. Glioblastoma: Defining Tumor Niches. *Trends in cancer* **1**, 252-265 (2015).

106.   Erdem-Eraslan, L., *et al.* Identification of Patients with Recurrent Glioblastoma Who May Benefit from Combined Bevacizumab and CCNU Therapy: A Report from the BELOB Trial. *Cancer research* **76**, 525-534 (2016).

107.   Grill, J., *et al.* Phase II, Open-Label, Randomized, Multicenter Trial (HERBY) of Bevacizumab in Pediatric Patients With Newly Diagnosed High-Grade Glioma. *Journal of clinical oncology : official journal of the American Society of Clinical Oncology* **36**, 951-958 (2018).

108.   Field, K.M., *et al.* Randomized phase 2 study of carboplatin and bevacizumab in recurrent glioblastoma. *Neuro-oncology* **17**, 1504-1513 (2015).

109.   Wick, W., *et al.* Lomustine and Bevacizumab in Progressive Glioblastoma. *The New England journal of medicine* **377**, 1954-1963 (2017).

110.   Mrdjen, D., *et al.* High-Dimensional Single-Cell Mapping of Central Nervous System Immune Cells Reveals Distinct Myeloid Subsets in Health, Aging, and Disease. *Immunity* **48**, 380-395 e386 (2018).

111.   Li, Q. & Barres, B.A. Microglia and macrophages in brain homeostasis and disease. *Nature reviews. Immunology* **18**, 225-242 (2018).

112.   Paolicelli, R.C., *et al.* Synaptic pruning by microglia is necessary for normal brain development. *Science* **333**, 1456-1458 (2011).

113.   Jordao, M.J.C., *et al.* Single-cell profiling identifies myeloid cell subsets with distinct fates during neuroinflammation. *Science* **363**(2019).

114.   Kierdorf, K., Masuda, T., Jordao, M.J.C. & Prinz, M. Macrophages at CNS interfaces: ontogeny and function in health and disease. *Nature reviews. Neuroscience* **20**, 547-562 (2019).

115.   Klemm, F., *et al.* Interrogation of the Microenvironmental Landscape in Brain Tumors Reveals Disease-Specific Alterations of Immune Cells. *Cell* **181**, 1643-1660 e1617 (2020).

116.   Bowman, R.L., *et al.* Macrophage Ontogeny Underlies Differences in Tumor-Specific Education in Brain Malignancies. *Cell reports* **17**, 2445-2459 (2016).

117.   Muller, S., *et al.* Single-cell profiling of human gliomas reveals macrophage ontogeny as a basis for regional differences in macrophage activation in the tumor microenvironment. *Genome biology* **18**, 234 (2017).

118.   Akkari, L., *et al.* Dynamic changes in glioma macrophage populations after radiotherapy reveal CSF-1R inhibition as a strategy to overcome resistance. *Sci Transl Med* **12**(2020).

119.   Brandenburg, S., *et al.* Resident microglia rather than peripheral macrophages promote vascularization in brain tumors and are source of alternative pro-angiogenic factors. *Acta neuropathologica* **131**, 365-378 (2016).

120.   Najafi, M., *et al.* Macrophage polarity in cancer: A review. *Journal of cellular biochemistry* **120**, 2756-2765 (2019).

121.   Gabrusiewicz, K., *et al.* Macrophage Ablation Reduces M2-Like Populations and Jeopardizes Tumor Growth in a MAFIA-Based Glioma Model. *Neoplasia* **17**, 374-384 (2015).

122.   Shi, Y., *et al.* Tumour-associated macrophages secrete pleiotrophin to promote PTPRZ1 signalling in glioblastoma stem cells for tumour growth. *Nature communications* **8**, 15080 (2017).

123.   Gabrusiewicz, K., *et al.* Glioblastoma-infiltrated innate immune cells resemble M0 macrophage phenotype. *JCI insight* **1**(2016).

124.   Szulzewsky, F., *et al.* Glioma-associated microglia/macrophages display an expression profile different from M1 and M2 polarization and highly express Gpnmb and Spp1. *PLoS One* **10**, e0116644 (2015).

125.   Chen, P., *et al.* Symbiotic Macrophage-Glioma Cell Interactions Reveal Synthetic Lethality in PTEN-Null Glioma. *Cancer cell* **35**, 868-884 e866 (2019).

126.   Wei, J., *et al.* Osteopontin mediates glioblastoma-associated macrophage infiltration and is a potential therapeutic target. *The Journal of clinical investigation* **129**, 137-149 (2019).

127.   Goswami, S., *et al.* Immune profiling of human tumors identifies CD73 as a combinatorial target in glioblastoma. *Nature medicine* **26**, 39-46 (2020).

128.   Gabrusiewicz, K., *et al.* Anti-vascular endothelial growth factor therapy-induced glioma invasion is associated with accumulation of Tie2-expressing monocytes. *Oncotarget* **5**, 2208-2220 (2014).

129.   Venneri, M.A., *et al.* Identification of proangiogenic TIE2-expressing monocytes (TEMs) in human peripheral blood and cancer. *Blood* **109**, 5276-5285 (2007).

130.   Scholz, A., *et al.* Endothelial cell-derived angiopoietin-2 is a therapeutic target in treatment-naive and bevacizumab-resistant glioblastoma. *EMBO molecular medicine* **8**, 39-57 (2016).

131.   Tamura, R., *et al.* Persistent restoration to the immunosupportive tumor microenvironment in glioblastoma by bevacizumab. *Cancer science* **110**, 499-508 (2019).

132.   Peterson, T.E., *et al.* Dual inhibition of Ang-2 and VEGF receptors normalizes tumor vasculature and prolongs survival in glioblastoma by altering macrophages. *Proceedings of the National Academy of Sciences of the United States of America* **113**, 4470-4475 (2016).

133.   Quail, D.F. & Joyce, J.A. The Microenvironmental Landscape of Brain Tumors. *Cancer cell* **31**, 326-341 (2017).

134.   Pyonteck, S.M., *et al.* CSF-1R inhibition alters macrophage polarization and blocks glioma progression. *Nature medicine* **19**, 1264-1272 (2013).

135.   Quail, D.F., *et al.* The tumor microenvironment underlies acquired resistance to CSF-1R inhibition in gliomas. *Science* **352**, aad3018 (2016).

136.   Yan, D., *et al.* Inhibition of colony stimulating factor-1 receptor abrogates microenvironment-mediated therapeutic resistance in gliomas. *Oncogene* **36**, 6049-6058 (2017).

137.   Butowski, N., *et al.* Orally administered colony stimulating factor 1 receptor inhibitor PLX3397 in recurrent glioblastoma: an Ivy Foundation Early Phase Clinical Trials Consortium phase II study. *Neuro-oncology* **18**, 557-564 (2016).

138.   Kaneda, M.M., *et al.* PI3Kgamma is a molecular switch that controls immune suppression. *Nature* **539**, 437-442 (2016).

139.   Guerriero, J.L., *et al.* Class IIa HDAC inhibition reduces breast tumours and metastases through anti-tumour macrophages. *Nature* **543**, 428-432 (2017).

140.   Liu, Y., *et al.* Abscopal effect of radiotherapy combined with immune checkpoint inhibitors. *Journal of hematology & oncology* **11**, 104 (2018).

141.   Lhuillier, C., *et al.* Radiotherapy-exposed CD8+ and CD4+ neoantigens enhance tumor control. *The Journal of clinical investigation* **131**(2021).

142.   Zeng, J., *et al.* Anti-PD-1 blockade and stereotactic radiation produce long-term survival in mice with intracranial gliomas. *International journal of radiation oncology, biology, physics* **86**, 343-349 (2013).

143.   Wang, S.M., *et al.* CCAAT/Enhancer-binding protein delta mediates glioma stem-like cell enrichment and ATP-binding cassette transporter ABCA1 activation for temozolomide resistance in glioblastoma. *Cell death discovery* **7**, 8 (2021).

144.   Bao, S., *et al.* Glioma stem cells promote radioresistance by preferential activation of the DNA damage response. *Nature* **444**, 756-760 (2006).

145.   Chen, J., *et al.* A restricted cell population propagates glioblastoma growth after chemotherapy. *Nature* **488**, 522-526 (2012).

146.   Birzu, C., *et al.* Recurrent Glioblastoma: From Molecular Landscape to New Treatment Perspectives. *Cancers* **13**(2020).

147.   Wang, K., *et al.* Receptor tyrosine kinase expression in high-grade gliomas before and after chemoradiotherapy. *Oncology letters* **18**, 6509-6515 (2019).

148.   Duan, C., *et al.* Late Effects of Radiation Prime the Brain Microenvironment for Accelerated Tumor Growth. *International journal of radiation oncology, biology, physics* **103**, 190-194 (2019).

149.   Berg, T.J., *et al.* The Irradiated Brain Microenvironment Supports Glioma Stemness and Survival via Astrocyte-Derived Transglutaminase 2. *Cancer research* **81**, 2101-2115 (2021).

150.   Roura, A.J., *et al.* Identification of the immune gene expression signature associated with recurrence of high-grade gliomas. *Journal of molecular medicine* **99**, 241-255 (2021).

151.   Zhang, P., *et al.* Therapeutic targeting of tumor-associated myeloid cells synergizes with radiation therapy for glioblastoma. *Proceedings of the National Academy of Sciences of the United States of America* **116**, 23714-23723 (2019).

152.   Thomas, R.P., *et al.* Macrophage Exclusion after Radiation Therapy (MERT): A First in Human Phase I/II Trial using a CXCR4 Inhibitor in Glioblastoma. *Clinical cancer research : an official journal of the American Association for Cancer Research* **25**, 6948-6957 (2019).

153.   Jeon, H.Y., *et al.* Ly6G(+) inflammatory cells enable the conversion of cancer cells to cancer stem cells in an irradiated glioblastoma model. *Cell death and differentiation* **26**, 2139-2156 (2019).

154.   Makarevic, A., *et al.* Increased Radiation-Associated T-Cell Infiltration in Recurrent IDH-Mutant Glioma. *International journal of molecular sciences* **21**(2020).

155.   Tamura, R., *et al.* Alterations of the tumor microenvironment in glioblastoma following radiation and temozolomide with or without bevacizumab. *Annals of translational medicine* **8**, 297 (2020).

156.   Seo, Y.S., *et al.* Radiation-Induced Changes in Tumor Vessels and Microenvironment Contribute to Therapeutic Resistance in Glioblastoma. *Frontiers in oncology* **9**, 1259 (2019).

157.   Pazzaglia, S., Briganti, G., Mancuso, M. & Saran, A. Neurocognitive Decline Following Radiotherapy: Mechanisms and Therapeutic Implications. *Cancers* **12**(2020).

158.   Kahalley, L.S., *et al.* Superior Intellectual Outcomes After Proton Radiotherapy Compared With Photon Radiotherapy for Pediatric Medulloblastoma. *Journal of clinical oncology : official journal of the American Society of Clinical Oncology* **38**, 454-461 (2020).

159.   Mohan, R., *et al.* Proton therapy reduces the likelihood of high-grade radiation-induced lymphopenia in glioblastoma patients: phase II randomized study of protons vs photons. *Neuro-oncology* **23**, 284-294 (2021).

160.   Petr, J., *et al.* Photon vs. proton radiochemotherapy: Effects on brain tissue volume and perfusion. *Radiotherapy and oncology : journal of the European Society for Therapeutic Radiology and Oncology* **128**, 121-127 (2018).

161.   Muroi, A., *et al.* Proton therapy for newly diagnosed pediatric diffuse intrinsic pontine glioma. *Child's nervous system : ChNS : official journal of the International Society for Pediatric Neurosurgery* **36**, 507-512 (2020).

162.   Jazmati, D., *et al.* Feasibility of Proton Beam Therapy for Infants with Brain Tumours: Experiences from the Prospective KiProReg Registry Study. *Clinical oncology* (2021).

163.   Scartoni, D., Amelio, D., Palumbo, P., Giacomelli, I. & Amichetti, M. Proton therapy re-irradiation preserves health-related quality of life in large recurrent glioblastoma. *Journal of cancer research and clinical oncology* **146**, 1615-1622 (2020).

164.   Atanasoski, S., *et al.* Cell cycle inhibitors p21 and p16 are required for the regulation of Schwann cell proliferation. *Glia* **53**, 147-157 (2006).

165.   Jessen, K.R. & Mirsky, R. The origin and development of glial cells in peripheral nerves. *Nature reviews. Neuroscience* **6**, 671-682 (2005).

166.   Laplante, M. & Sabatini, D.M. mTOR signaling at a glance. *Journal of cell science* **122**, 3589-3594 (2009).

167.   Inoki, K., Li, Y., Zhu, T., Wu, J. & Guan, K.L. TSC2 is phosphorylated and inhibited by Akt and suppresses mTOR signalling. *Nature cell biology* **4**, 648-657 (2002).

168.   Hay, N. & Sonenberg, N. Upstream and downstream of mTOR. *Genes & development* **18**, 1926-1945 (2004).

169.   Huang, J. & Manning, B.D. The TSC1-TSC2 complex: a molecular switchboard controlling cell growth. *The Biochemical journal* **412**, 179-190 (2008).

170.   D'Antonio, M., *et al.* TGFbeta type II receptor signaling controls Schwann cell death and proliferation in developing nerves. *The Journal of neuroscience : the official journal of the Society for Neuroscience* **26**, 8417-8427 (2006).

171.   Deng, Y., *et al.* A reciprocal regulatory loop between TAZ/YAP and G-protein Galphas regulates Schwann cell proliferation and myelination. *Nature communications* **8**, 15161 (2017).

172.   Ghidinelli, M., *et al.* Laminin 211 inhibits protein kinase A in Schwann cells to modulate neuregulin 1 type III-driven myelination. *PLoS biology* **15**, e2001408 (2017).

173.   Herbert, A.L. & Monk, K.R. Advances in myelinating glial cell development. *Current opinion in neurobiology* **42**, 53-60 (2017).

174.   Ma, Z., Wang, J., Song, F. & Loeb, J.A. Critical period of axoglial signaling between neuregulin-1 and brain-derived neurotrophic factor required for early Schwann cell survival and differentiation. *The Journal of neuroscience : the official journal of the Society for Neuroscience* **31**, 9630-9640 (2011).

175.   Maurel, P. & Salzer, J.L. Axonal regulation of Schwann cell proliferation and survival and the initial events of myelination requires PI 3-kinase activity. *The Journal of neuroscience : the official journal of the Society for Neuroscience* **20**, 4635-4645 (2000).

176.   Yamauchi, J., Miyamoto, Y., Chan, J.R. & Tanoue, A. ErbB2 directly activates the exchange factor Dock7 to promote Schwann cell migration. *The Journal of cell biology* **181**, 351-365 (2008).

177.   Porrello, E., *et al.* Jab1 regulates Schwann cell proliferation and axonal sorting through p27. *The Journal of experimental medicine* **211**, 29-43 (2014).

178.   Tikoo, R., Zanazzi, G., Shiffman, D., Salzer, J. & Chao, M.V. Cell cycle control of Schwann cell proliferation: role of cyclin-dependent kinase-2. *The Journal of neuroscience : the official journal of the Society for Neuroscience* **20**, 4627-4634 (2000).

179.   Nakashima, A., *et al.* The yeast Tor signaling pathway is involved in G2/M transition via polo-kinase. *PloS one* **3**, e2223 (2008).

180.   Renner, A.G., *et al.* A functional link between polo-like kinase 1 and the mammalian target-of-rapamycin pathway? *Cell cycle* **9**, 1690-1696 (2010).

181.   Barr, F.A., Sillje, H.H. & Nigg, E.A. Polo-like kinases and the orchestration of cell division. *Nature reviews. Molecular cell biology* **5**, 429-440 (2004).

182.   Liu, X. & Erikson, R.L. Activation of Cdc2/cyclin B and inhibition of centrosome amplification in cells depleted of Plk1 by siRNA. *Proceedings of the National Academy of Sciences of the United States of America* **99**, 8672-8676 (2002).

183.   Tang, J., Erikson, R.L. & Liu, X. Ectopic expression of Plk1 leads to activation of the spindle checkpoint. *Cell cycle* **5**, 2484-2488 (2006).

184.   Beirowski, B., Wong, K.M., Babetto, E. & Milbrandt, J. mTORC1 promotes proliferation of immature Schwann cells and myelin growth of differentiated Schwann cells. *Proceedings of the National Academy of Sciences of the United States of America* **114**, E4261-E4270 (2017).

185.   Bercury, K.K., *et al.* Conditional ablation of raptor or rictor has differential impact on oligodendrocyte differentiation and CNS myelination. *The Journal of neuroscience : the official journal of the Society for Neuroscience* **34**, 4466-4480 (2014).

186.   Figlia, G., Gerber, D. & Suter, U. Myelination and mTOR. *Glia* **66**, 693-707 (2018).

187.   Lebrun-Julien, F., *et al.* Balanced mTORC1 activity in oligodendrocytes is required for accurate CNS myelination. *The Journal of neuroscience : the official journal of the Society for Neuroscience* **34**, 8432-8448 (2014).

188.   Wahl, S.E., McLane, L.E., Bercury, K.K., Macklin, W.B. & Wood, T.L. Mammalian target of rapamycin promotes oligodendrocyte differentiation, initiation and extent of CNS myelination. *The Journal of neuroscience : the official journal of the Society for Neuroscience* **34**, 4453-4465 (2014).

189.   Figlia, G., Norrmen, C., Pereira, J.A., Gerber, D. & Suter, U. Dual function of the PI3K-Akt-mTORC1 axis in myelination of the peripheral nervous system. *eLife* **6**(2017).

190.   Valdivia, A.O., Farr, V. & Bhattacharya, S.K. A novel myelin basic protein transcript variant in the murine central nervous system. *Molecular biology reports* **46**, 2547-2553 (2019).

191.   Flores, A.I., *et al.* Constitutively active Akt induces enhanced myelination in the CNS. *The Journal of neuroscience : the official journal of the Society for Neuroscience* **28**, 7174-7183 (2008).

192.   Hsieh, H.J., *et al.* Systems biology approach reveals a link between mTORC1 and G2/M DNA damage checkpoint recovery. *Nature communications* **9**, 3982 (2018).

193.   Dehay, C. & Kennedy, H. Cell-cycle control and cortical development. *Nature reviews. Neuroscience* **8**, 438-450 (2007).

194.   Zhang, S., *et al.* B cell-specific deficiencies in mTOR limit humoral immune responses. *Journal of immunology* **191**, 1692-1703 (2013).

195.   Zhang, S., *et al.* Constitutive reductions in mTOR alter cell size, immune cell development, and antibody production. *Blood* **117**, 1228-1238 (2011).

196.   Tricaud, N. Myelinating Schwann Cell Polarity and Mechanically-Driven Myelin Sheath Elongation. *Frontiers in cellular neuroscience* **11**, 414 (2017).

197.   Zorick, T.S., Syroid, D.E., Arroyo, E., Scherer, S.S. & Lemke, G. The Transcription Factors SCIP and Krox-20 Mark Distinct Stages and Cell Fates in Schwann Cell Differentiation. *Molecular and cellular neurosciences* **8**, 129-145 (1996).

198.   Stupp, R., *et al.* Radiotherapy plus concomitant and adjuvant temozolomide for glioblastoma. *The New England journal of medicine* **352**, 987-996 (2005).

199.   Sa, J.K., *et al.* Transcriptional regulatory networks of tumor-associated macrophages that drive malignancy in mesenchymal glioblastoma. *Genome biology* **21**, 216 (2020).

200.   Flores-Toro, J.A., *et al.* CCR2 inhibition reduces tumor myeloid cells and unmasks a checkpoint inhibitor effect to slow progression of resistant murine gliomas. *Proceedings of the National Academy of Sciences of the United States of America* **117**, 1129-1138 (2020).

201.   Coniglio, S.J., *et al.* Microglial stimulation of glioblastoma invasion involves epidermal growth factor receptor (EGFR) and colony stimulating factor 1 receptor (CSF-1R) signaling. *Molecular medicine* **18**, 519-527 (2012).

202.   Pombo Antunes, A.R., *et al.* Single-cell profiling of myeloid cells in glioblastoma across species and disease stage reveals macrophage competition and specialization. *Nature neuroscience* **24**, 595-610 (2021).

203.   Zudaire, E., Gambardella, L., Kurcz, C. & Vermeren, S. A computational tool for quantitative analysis of vascular networks. *PloS one* **6**, e27385 (2011).

204.   Andrews, S. FastQC: A quality control tool for high throughput sequence data. (2010).

205.   Krueger, F. A wrapper tool around Cutadapt and FastQC to consistently apply quality and adapter trimming to FastQ files, with some extra functionality for MspI-digested RRBS-type (Reduced Representation Bisufite-Seq) libraries.

206.   Martin, M. Cutadapt removes adapter sequences from high-throughput sequencing reads. *EMBnet.journal* **17**, 10-12 (2011).

207.   Dobin, A., *et al.* STAR: ultrafast universal RNA-seq aligner. *Bioinformatics* **29**, 15-21 (2013).

208.   Tarasov, A., Vilella, A.J., Cuppen, E., Nijman, I.J. & Prins, P. Sambamba: fast processing of NGS alignment formats. *Bioinformatics* **31**, 2032-2034 (2015).

209.   O'Leary, N.A., *et al.* Reference sequence (RefSeq) database at NCBI: current status, taxonomic expansion, and functional annotation. *Nucleic acids research* **44**, D733-745 (2016).

210.   Liao, Y., Smyth, G.K. & Shi, W. The R package Rsubread is easier, faster, cheaper and better for alignment and quantification of RNA sequencing reads. *Nucleic acids research* **47**, e47 (2019).

211.   Korsunsky, I., *et al.* Fast, sensitive and accurate integration of single-cell data with Harmony. *Nature methods* **16**, 1289-1296 (2019).

212.   Korsunsky, I., Nathan, A., Millard, N. & Raychaudhuri, S. Presto scales Wilcoxon and auROC analyses to millions of observations. *bioRxiv*, 653253 (2019).

213.   Wu, L.M.N., *et al.* Programming of Schwann Cells by Lats1/2-TAZ/YAP Signaling Drives Malignant Peripheral Nerve Sheath Tumorigenesis. *Cancer cell* **33**, 292-308 e297 (2018).

214.   Jin, S., *et al.* Inference and analysis of cell-cell communication using CellChat. *Nature communications* **12**, 1088 (2021).

215.   Yuan, J., *et al.* Single-cell transcriptome analysis of lineage diversity in high-grade glioma. *Genome medicine* **10**, 57 (2018).

216.   Darmanis, S., *et al.* Single-Cell RNA-Seq Analysis of Infiltrating Neoplastic Cells at the Migrating Front of Human Glioblastoma. *Cell reports* **21**, 1399-1410 (2017).

217.   Weng, Q., *et al.* Single-Cell Transcriptomics Uncovers Glial Progenitor Diversity and Cell Fate Determinants during Development and Gliomagenesis. *Cell Stem Cell* **24**, 707-723 e708 (2019).

218.   Zhang, S., *et al.* m(6)A Demethylase ALKBH5 Maintains Tumorigenicity of Glioblastoma Stem-like Cells by Sustaining FOXM1 Expression and Cell Proliferation Program. *Cancer cell* **31**, 591-606 e596 (2017).

219.   Wu, S., *et al.* EGFR Amplification Induces Increased DNA Damage Response and Renders Selective Sensitivity to Talazoparib (PARP Inhibitor) in Glioblastoma. *Clinical cancer research : an official journal of the American Association for Cancer Research* **26**, 1395-1407 (2020).

220.   Friedmann-Morvinski, D., *et al.* Targeting NF-kappaB in glioblastoma: A therapeutic approach. *Science advances* **2**, e1501292 (2016).

221.   Patel, S.K., *et al.* Generation of diffuse intrinsic pontine glioma mouse models by brainstem-targeted in utero electroporation. *Neuro-oncology* **22**, 381-392 (2020).

222.   Saederup, N., *et al.* Selective chemokine receptor usage by central nervous system myeloid cells in CCR2-red fluorescent protein knock-in mice. *PloS one* **5**, e13693 (2010).

223. Macosko, E.Z., *et al.* Highly Parallel Genome-wide Expression Profiling of Individual Cells Using Nanoliter Droplets. *Cell* **161**, 1202-1214 (2015).

224. Hammond, T.R., *et al.* Single-Cell RNA Sequencing of Microglia throughout the Mouse Lifespan and in the Injured Brain Reveals Complex Cell-State Changes. *Immunity* **50**, 253-271 e256 (2019).

225. de Groot, J., *et al.* Modulating antiangiogenic resistance by inhibiting the signal transducer and activator of transcription 3 pathway in glioblastoma. *Oncotarget* **3**, 1036-1048 (2012).

226. Lobo, M.R., *et al.* Synergistic Antivascular and Antitumor Efficacy with Combined Cediranib and SC6889 in Intracranial Mouse Glioma. *PLoS One* **10**, e0144488 (2015).

227. Ding, P., Wang, W., Wang, J., Yang, Z. & Xue, L. Expression of tumor-associated macrophage in progression of human glioma. *Cell biochemistry and biophysics* **70**, 1625-1631 (2014).

228. Ozawa, T., *et al.* Most human non-GCIMP glioblastoma subtypes evolve from a common proneural-like precursor glioma. *Cancer Cell* **26**, 288-300 (2014).

229. von Roemeling, C.A., *et al.* Therapeutic modulation of phagocytosis in glioblastoma can activate both innate and adaptive antitumour immunity. *Nat Commun* **11**, 1508 (2020).

230. Filbin, M.G., *et al.* Developmental and oncogenic programs in H3K27M gliomas dissected by single-cell RNA-seq. *Science* **360**, 331-335 (2018).

231. Becher, O.J., *et al.* Preclinical evaluation of radiation and perifosine in a genetically and histologically accurate model of brainstem glioma. *Cancer research* **70**, 2548-2557 (2010).

232. Zhang, L., *et al.* Pleiotrophin promotes vascular abnormalization in gliomas and correlates with poor survival in patients with astrocytomas. *Science signaling* **8**, ra125 (2015).

233. Vasse, M., *et al.* Oncostatin M induces angiogenesis in vitro and in vivo. *Arteriosclerosis, thrombosis, and vascular biology* **19**, 1835-1842 (1999).

234. Monney, L., *et al.* Th1-specific cell surface protein Tim-3 regulates macrophage activation and severity of an autoimmune disease. *Nature* **415**, 536-541 (2002).

235. Kijewska, M., *et al.* The embryonic type of SPP1 transcriptional regulation is re-activated in glioblastoma. *Oncotarget* **8**, 16340-16355 (2017).

236. Pietras, A., *et al.* Osteopontin-CD44 signaling in the glioma perivascular niche enhances cancer stem cell phenotypes and promotes aggressive tumor growth. *Cell stem cell* **14**, 357-369 (2014).

237. Xu, L., *et al.* BCL6 promotes glioma and serves as a therapeutic target. *Proceedings of the National Academy of Sciences of the United States of America* **114**, 3981-3986 (2017).

238. Guo, G., *et al.* A TNF-JNK-Axl-ERK signaling axis mediates primary resistance to EGFR inhibition in glioblastoma. *Nature neuroscience* **20**, 1074-1084 (2017).

239. Lu, F., *et al.* Olig2-Dependent Reciprocal Shift in PDGF and EGF Receptor Signaling Regulates Tumor Phenotype and Mitotic Growth in Malignant Glioma. *Cancer Cell* **29**, 669-683 (2016).

240. Wang, Z., *et al.* Cell Lineage-Based Stratification for Glioblastoma. *Cancer cell* **38**, 366-379 e368 (2020).

241. Oweida, A.J., *et al.* Response to radiotherapy in pancreatic ductal adenocarcinoma is enhanced by inhibition of myeloid-derived suppressor cells using STAT3 anti-sense oligonucleotide. *Cancer immunology, immunotherapy : CII* **70**, 989-1000 (2021).

242. Di Mitri, D., *et al.* Re-education of Tumor-Associated Macrophages by CXCR2 Blockade Drives Senescence and Tumor Inhibition in Advanced Prostate Cancer. *Cell reports* **28**, 2156-2168 e2155 (2019).

243. Blank, A., *et al.* Microglia/macrophages express alternative proangiogenic factors depending on granulocyte content in human glioblastoma. *The Journal of pathology* **253**, 160-173 (2021).

244. Hennig, J., Nauerth, A. & Friedburg, H. RARE imaging: a fast imaging method for clinical MR. *Magnetic resonance in medicine* **3**, 823-833 (1986).

245. Kolberg, M., *et al.* Drug sensitivity and resistance testing identifies PLK1 inhibitors and gemcitabine as potent drugs for malignant peripheral nerve sheath tumors. *Molecular oncology* **11**, 1156-1171 (2017).

246. Huang, Z., Powell, R., Phillips, J.B. & Haastert-Talini, K. Perspective on Schwann Cells Derived from Induced Pluripotent Stem Cells in Peripheral Nerve Tissue Engineering. *Cells* **9**(2020).

247. Abramsson, A., Lindblom, P. & Betsholtz, C. Endothelial and nonendothelial sources of PDGF-B regulate pericyte recruitment and influence vascular pattern formation in tumors. *The Journal of clinical investigation* **112**, 1142-1151 (2003).

248. Hasselmann, J., *et al.* Development of a Chimeric Model to Study and Manipulate Human Microglia In Vivo. *Neuron* **103**, 1016-1033 e1010 (2019).

249. Rojo, R., *et al.* Deletion of a Csf1r enhancer selectively impacts CSF1R expression and development of tissue macrophage populations. *Nature communications* **10**, 3215 (2019).

250. Chiblak, S., *et al.* Carbon irradiation overcomes glioma radioresistance by eradicating stem cells and forming an antiangiogenic and immunopermissive niche. *JCI insight* **4**(2019).

251. Oh, T., *et al.* Immunocompetent murine models for the study of glioblastoma immunotherapy. *Journal of translational medicine* **12**, 107 (2014).

252. Zhang, L., *et al.* Single-Cell Transcriptomics in Medulloblastoma Reveals Tumor-Initiating Progenitors and Oncogenic Cascades during Tumorigenesis and Relapse. *Cancer cell* **36**, 302-318 e307 (2019).

253. Veglia, F., Perego, M. & Gabrilovich, D. Myeloid-derived suppressor cells coming of age. *Nature immunology* **19**, 108-119 (2018).

254. Alban, T.J., *et al.* Global immune fingerprinting in glioblastoma patient peripheral blood reveals immune-suppression signatures associated with prognosis. *JCI insight* **3**(2018).

255. Raychaudhuri, B., *et al.* Myeloid derived suppressor cell infiltration of murine and human gliomas is associated with reduction of tumor infiltrating lymphocytes. *Journal of neuro-oncology* **122**, 293-301 (2015).

256. Armstrong, C.W.D., *et al.* Clinical and functional characterization of CXCR1/CXCR2 biology in the relapse and radiotherapy resistance of primary PTEN-deficient prostate carcinoma. *NAR cancer* **2**, zcaa012 (2020).

257. Peereboom, D.M., *et al.* Metronomic capecitabine as an immune modulator in glioblastoma patients reduces myeloid-derived suppressor cells. *JCI insight* **4**(2019).

258. Damo, M., *et al.* Inducible de novo expression of neoantigens in tumor cells and mice. *Nature biotechnology* **39**, 64-73 (2021).